GAIA
A Way of Knowing

GAIA
A Way of Knowing

Political Implications of the New Biology

edited by

WILLIAM IRWIN THOMPSON

LINDISFARNE PRESS

GAIA: A WAY OF KNOWING
IS A PUBLICATION OF THE LINDISFARNE PROGRAM
FOR BIOLOGY, COGNITION AND ETHICS, THE CATHEDRAL CHURCH
OF ST. JOHN THE DIVINE, NEW YORK CITY, MADE POSSIBLE BY
GRANTS FROM TRINITY CHURCH, NEW YORK CITY,
AND THE PRINCE CHARITABLE TRUSTS, CHICAGO.

"GAIA AND THE POLITICS OF LIFE: A PROGRAM FOR THE NINETIES"
ORIGINALLY APPEARED IN SLIGHTLY DIFFERENT FORM AS
"A GAIAN POLITICS" IN *WHOLE EARTH REVIEW*, No. 53, WINTER, 1986.

PART OF "GAIA: A MODEL FOR PLANETARY AND CELLULAR
DYNAMICS" ORIGINALLY APPEARED AS "MORE ON GAIA"
IN *COEVOLUTION QUARTERLY*, NO. 31, FALL, 1981.

10 9 8 7 6 5

PUBLISHED BY THE LINDISFARNE PRESS,
195 MAIN STREET, GREAT BARRINGTON, MA 01230.

LIBRARY OF CONGRESS CATALOGING-IN-PUBLICATION DATA
GAIA, A WAY OF KNOWING.

CHIEFLY PAPERS PRESENTED AT LINDISFARNE FELLOWS
CONFERENCE, 1981, SAN FRANCISCO ZEN CENTER,
SAN FRANCISCO CALIF.; SPONSORED BY LINDISFARNE
ASSOCIATION.
INCLUDES BIBLIOGRAPHIES.
1. BIOLOGY—PHILOSOPHY—CONGRESSES. 2. BIOLOGY—
SOCIAL ASPECTS—CONGRESSES. I. THOMPSON, WILLIAM
IRWIN. II. LINDISFARNE FELLOWS CONFERENCE (1981 :
SAN FRANCISCO ZEN CENTER) III. LINDISFARNE ASSOCIATION.
QH331.G22 1987 574'.01 87-10112
ISBN 0-940262-23-1 (PBK.)

PRINTED IN THE UNITED STATES OF AMERICA

CONTENTS

PART TWO
Gaia Politique

PREFACE

IDEAS, LIKE GRAPES, grow in clusters. People like to hang out together because they can feel their ideas growing fuller and richer on the vine. This book is just such a cluster of ideas that comes from a small group of people who have been hanging out together for the last six years. More than anything it is a work of intellectual fellowship that expresses the ideas, conferences, conversations, letters, and phone calls that have been going on since each person began to recognize that there were dimensions of his or her own work that did not show up when one looked into the mirror built into the vanity of one's private work, but did appear when one saw one's work described and extended in the ideas of a friend.

We came together as a group for the first time in 1981; it was at my invitation to one of the small Lindisfarne Fellows conferences that I have been holding for the last twelve years or so. It was a good conference, not necessarily better than all the others, but more powerful in the sense that it took over my own work and began deflecting the stream of my thoughts from anthropology and prehistory into biology. As I read Lovelock's *Gaia: A New Look at Life on Earth*, or looked at Lynn Margulis's film about bacteria, my thoughts on the relationship between myth and science took a jump forward as I began to appreciate just what my Irish ancestors had meant when they talked about "the little people" at work in the leaf mold at their feet. Lynn talked of bacteria laying down the iron ore deposits in the Gunflint formations of Ontario, and I saw dwarves at work in the mines.

Imagination is needed to shape a theory or a hypothesis, and Whitehead argued a long time ago that pure induction could never produce a scientific view of the world. A heap of facts was useless, and neither a Homeric epic nor a scientific theory of evolution could ever be produced from mere facts. For people in a prescientific culture, people endowed with acute powers of observation and remarkable sensitivity, there was no way to imagine the life at one's feet except through the poetic imagination which made the little creatures half human. And in a way, this poetic imagination of the ancients is more sensitive to humanity's embeddedness in the biosphere, for in seeing "the little people" as half human the ancient Irish "fairy faith" recognized that there is no "us" and "them," that we are in them, and they are in us.

The imagination is, therefore, not a source of deception and delusion, but a capacity to sense what you do not know, to intuit what you cannot understand, to *be* more than you can *know*. The imaging capacity of the mind is not the epiphenomenal discharge of purer processes of computational logic, as today's reductionists in Artificial Intelligence would have us believe; no, the image is a transform of awareness from other dimensions of sensitivity. The song you did not hear you may begin to hum. The bacteria you did not see you may begin to envision.

This capacity to think in images, and then transform them into other dimensions of reference is vital to art, poetry, and science. High on a peak in the Andes, Darwin's imagination took fire as he saw how "we are all netted together." Raging in a tropical fever, Alfred Russell Wallace saw the lineaments of the theory of natural selection. Whether in Descartes' dream of philosophy or in Kekulé's dream of the benzine ring, the image is not an imprecise and prescientific way of thinking. Facts depend on theories just as much as theories depend on facts. When raised on high, "facts" can become superstitious idols with which an imperial power strives to wipe out a traditional culture. If one speaks, in the art of *Feng-shui,* of the dragons in the sky, or the dragon current in the soil, one is being far more scientific than the modern chemist of agribusiness who destroys the soil and ruins the water table. An image is not an illusion; it is a hieroglyph, a compressed story, and stories are literally forms of cultural storage. Rather than imagery and poetry being the simple stuff of fairy tales, and facts being the real stuff of "hard science," it is hard facts that are simple-minded and imagery that is dazzlingly complex. An image is more like a historical fugue filled with references to old melodies as it transforms them into new variations. And just as one can be moved by Bach's *Art of the Fugue* without understanding his use of Leibniz's *Ars combinatoria,* so one can be moved by a myth or a fairy tale without an intellectual understanding of it. Precisely because we are more than we know, science can never embrace the totality of Being. At the edge of our knowing is a horizon, a place where clouds come out of the sky and move over our island home. This process in which a vast planetary atmosphere condenses into the distinct shape of a cloud is also the process in which a vast and global form of knowledge becomes a distinct image in our limited personality, the ego. And in this metaphor of the atmosphere and the cloud is an imagistic performance of the very philosophy I am seeking to describe.

Performance is the right word for it, for what I am offering in this book is not so much a description of some scientific theories but an unfoldment in which the observer of the scientific observer changes the science of the scientist. The literary writer, the poet, becomes possessed by science, and in reflecting the work back to the scientist, the scientist sees his image transformed. Over his shoulder in the mirror he sees himself involved in a cultural landscape he had not noticed before. He or she sees the mythic structures of the imagination and discovers that science and the humanities are moving into a postmodernist world in which neither one is what it was before. This is not a case of an easily split world, with soft subjectivity to one side and hard objectivity on the other, with humanities here and science there; it is a new condition of biology and the way of knowing.

Before Gaia was a hypothesis she was a goddess, so what more appropriate area could there be for an exploration of myth and science? But the Gaia hypothesis alone would not be enough to express the way of knowing or the politics of life. With the atmospheric chemistry of Lovelock, we have the macrocosm; with the bacteriology of Margulis we have the microcosm, but moving between the macrocosm of the planet and the microcosm of the cell is the mesocosm of the mind. It is here in the cognitive biology of Maturana and Varela that knowing truly becomes the organization of the living that brings forth a world.

William Irwin Thompson
Cathedral of St. John the Divine
New York City
February, 1987

INTRODUCTION

WILLIAM IRWIN THOMPSON

The Cultural Implications of the New Biology

THE PURPOSE OF the Lindisfarne Conferences, from my perspective as host and organizer, is to render explicit an implicit harmony that exists in the works of those who take part in these gatherings. Some of you are meeting one another for the first time, and this kind of meeting is, I feel, a very important aspect of the work of Lindisfarne. In my travels I meet people whom I recognize as fellow-workers in the emergence of a new culture, and I sense a need that they should meet someone who may be unknown to them personally, but is, nevertheless, a colleague of theirs in the common work. This pattern of association has grown over the years to become the very non-institutional group known as the Lindisfarne Fellows.

This gathering is very much a continuation of that spirit of intellectual and spiritual fellowship. We meet in the room here where last year we heard the tape of what was to be Gregory Bateson's farewell address to the Fellows. Certainly this is a gathering that Gregory would have deeply enjoyed, for present here tonight are some of the individuals, going all the way back to the original Macy Conferences, who have been responsible for opening up new paths in cybernetics, epistemology, and self-organizing systems biology.

To honor truly the work that Bateson began in the thirties we must not simply honor his memory, but must continue to take further steps toward an ecology of mind. "The pattern that connects" all of us gathered here tonight is, as I see it, as follows. Flowing out of the work of both Gregory Bateson, Heinz Von Foerster, and connecting with the original work of Warren McCulloch at M.I.T. is the Santiago school of cognitive biology, expressed in the work of Humberto Maturana and Francisco Varela. Stimulated by this early work in bringing together biology and information theory is what I shall call the Parisian school of self-organizing systems biology, represented here by Henri Atlan. Now there comes along a quite separate stream which has its source in a different mountaintop, and this stream is the work of James Lovelock and Lynn Margulis in the formulation of the Gaia hypothesis as a model for planetary and cellular dynamics. Now I would propose that if we fly up to a higher level and look down on this intellectual landscape, as with the eye of a Landsat camera, we can see that all these streams are involved in a common watershed, and that all are flowing into and nourishing a single lake. I see this lake as a metaphor for a new polity, a new ecology of consciousness. As these different streams of thought begin to connect in a larger ecosystem, we begin to move out from the particular performance of scientific research and innovation to the idea of not simply a new finding or a new theory, but a new planetary culture. Such is the open agenda of Lindisfarne.

And now let me return to my particular work, not as host, but as cultural historian. One of the things that fascinates me about science as a cultural activity is the way in which scientific *narratives* (and here I am purposely taking the word from fiction) are rooted in unconscious conceptions of order. Even when a scientist thinks he is being meticulously rational, when he begins to take his data and organize them into a narrative, the facts become like morphemes organized by a grammar into a language of descriptions. These descriptions of "reality" or "nature" are the narratives I am referring to. The scientist cannot help but inherit the narrative traditions of his culture, of those forms of connecting events that may be as simple as Beginning, Middle, and End, or more complex four-fold patterns as expressed in Carnot cycles or Viconian cycles of history, or artistic styles that move from Archaic to Classic to Baroque to Archaistic.

All narratives, artistic, historical, or scientific, are connected to certain unconscious principles of ordering both our perceptions and our descriptions. For example, it is a truism of scholarship to hail Thucydides' *Peloponnesian Wars* as the first true work of scientific, as opposed to mythic and legendary, history. And yet when we look closely at his description of the Syracusan expedition, of the sailing out of the Athenian armada to its fated disaster, we notice that the narrative is analogous to the description of Patroklos donning the armor of Achilles and going out beyond the wall, beyond the limit, to his disaster. In Homeric epic tragedy, one's unique excellence (*aretê*) is inseparable from his tragic flaw (*hamartia*). Under Themistocles, the Greeks were daring risk takers and this led to their victory over the Persians at Salamis. And so when Alcibiades is being the daring risk taker in the Syracusan expedition, he is being quintessentially Athenian. But the *kairos* is wrong; the limit has been passed, and so the nature that creates glorious victory for Themistocles creates ignominious defeat for Alcibiades. My point is that when Thucydides tries to move beyond mythic history to write scientific history, he still sees history in the forms of the mythopoeic structures he has inherited from the Homeric epic. I, of course, see this isomorphism not as a sign of weakness but as a source of strength. History, wrought to its uttermost, achieves the condition of the unique excellence of poetry. A mindless and mechanical history of accumulating facts tells us very little, but a mindful history reveals the universal truth of events, as Aristotle pointed out a long time ago.

All narratives are structurings of time and are therefore inescapably related to unconscious systems of ordering. From the sociology of knowledge created by Feuerbach and Marx we have learned how to relate these narratives to the economic and political conditions of a particular culture. I would accept their insights but insist we go even deeper. The organization of knowledge and the organization of society are both related to deeper levels of the organization of perception and consciousness. You can be Kantian and see these as "pure concepts of the understanding," or Platonic to see these as archetypal forms of the Intelligible world that determine the phenomena of the Sensible world. I lean a little more to the Platonic side and see these narratives as shaped by archetypal ideas of order. But I have also been taught by our Zen Buddhist colleagues present here tonight that these Platonic ideas should not be reified in some heavenly empyrean, that they too are involved in a "codependent origination" that is empty of all absolutistic solidity. All things melt into *sunyata,* but not at the same temperature. An object melts quite quickly, but an archetype with which we structure the perception of objects melts more slowly and at a much higher temperature. So, as you can see, I am Marxist, Platonic, and Buddhist, a true child of the last quarter of the twentieth century.

Now that I have been so philosophical and general, let me be more specific and particular to give you some examples of scientific narratives that are rooted in unconscious ideas of order. Here is Darwin from *The Origin of Species*:

> Natural selection can produce nothing in one
> species for the exclusive good or injury of

> another; though it may produce parts, organs, and excretions highly useful or even indispensable, or again highly injurious to another species, but in all cases at the same time useful to the possessor. In each well-stocked country natural selection acts through the competition of the inhabitants, and consequently leads to success in the battle for life, only in accordance with the standard of the particular country. Hence the inhabitants of one country, generally the smaller one, often yield to the inhabitants of another and generally the larger country. For in the larger country there will have existed more individuals and more diversified forms, and the competition will have been severer, and thus the standard of perfection will have been rendered higher.[1]

Here the system of ordering is seen in the form of a relationship between a dominant and a subordinate group, and what is being presented in the biological narrative is the political relationship between England and Ireland. Social Darwinism is not a later perversion of Darwinism; Social Darwinism is Darwinism; the organization of knowledge and the organization of society derive from a single externalization of consciousness, a single *zeitgeist.* England first rejected Darwinism, or let us say that the clergy and the traditional landowning class rejected the unconscious apologetics of the new liberal business class, but when by the time of the 1880s the new capitalist class had firmly consolidated its power through the Industrial Revolution, the popular London *Illustrated News* took Darwinism to its heart and began publishing cartoons in which the Irish were drawn with simian features. The Irishman became the missing link, and the Englishman became the helmsman of planetary evolution. Darwinism as the political apologetics for the British Empire could not be more clearly expressed than in the rhapsodic finale to *The Origin of Species*:

> We can so far take a prophetic glance into futurity as to foretell that it will be the common and widely-spread species, belonging to the larger and dominant groups within each class, which will ultimately prevail and procreate new and dominant species. As all the living forms of life are the lineal descendants of those which lived long before the Cambrian epoch, we may feel certain that the ordinary succession by generation has never once been broken, and that no cataclysm has desolated the whole world. Hence we may look with some confidence to a secure future of great length. And as natural selection works solely by and for the good of each being, all corporeal and mental endowments will tend to progress towards perfection.[2]

This kind of scientific narrative is really a description of the British Empire, "the common and widely-spread species, belonging to the larger and dominant groups," and is a celebration of the rationale for the dominance of that group in the mythic notion of progress. It is rather interesting to observe how the celebration of the concept of progress does seem to require a rejection of catastrophism as being part of the dynamics of change and/or development. In the nineteenth century, especially in the work of the paleontologist Cuvier, catastrophism was the dominant idea, and it was an idea that appealed to people of religious orientation, for it fit in with notions of the Fall and the wrath of Jahweh. When Hutton and Lyell succeeded in replacing catastrophism with uniformitarianism they provided Darwin with the vast stretches of time he needed for his model of development through natural selection to take place. For Lyell nature behaved like an English gentleman; there were no vulgar and sudden upheavals to disturb the natural order of progress toward perfection through science and reason. And so the concept of progress established the metaphysical foundations for industrial society.

Today, however, the picture is quite different. Catastrophism is returning, and so we must begin to ask ourselves whether the metaphysical foundations of industrial society are also crumbling away. There is much talk about the Cretaceous extinction, and certainly James Lovelock has done as much as anyone to present a picture of catastrophe as part of the dynamics of planetary evolution. In *Gaia* we begin with the evolution of the solar system as the result of a supernova or the collapse of a binary star; then we move into a closer focus on Earth itself to consider that one of the greatest catastrophes that occurred was the shift to an aerobic atmosphere, when oxygen became a universal poison for all the well-established anaerobic life on Earth. And from there we can move down the line of time to the catastrophe of the Cretaceous extinction. Just as the forms of Homeric tragedy were the organizing principles for the narratives of Thucydides, so the idea of catastrophe is the organizing principle for Lovelock's narrative of Gaia.

Lest we try to blame all the bad news on our welcome guest, James Lovelock, it is only fair to point out that his is only one of the more scientific presentations of the idea, and that the idea of catastrophism has also found its way into the public imagination through art. We see a vision of catastrophe in the films of Peter Weir and Werner Herzog, in the recent space-fiction novels of Doris Lessing, in the *Book of the Hopi*, or the prophecies of native American leaders like Philip Deer. And just about a year ago, in this very room, His Holiness the Dalai Lama, made an observation that sounded like straight Gregory Bateson. His Holiness said: "When man changes the environment at too rapid a rate, say, for example, by turning the oceans of oil in the earth's crust into a gas in the earth's atmosphere, he creates a situation in which the environment changes faster than his own rate of adaption." In 1976 the Venerable Nechung Rinpoche shared with us at Lindisfarne some of the prophecies of the Oracle of Tibet, as well as some of the more ancient prophecies, and so the cultural milieu out of which the Dalai Lama was speaking is not what we would call "scientific," and yet one could not ask for a more crisp presentation of catastrophe as "a discontinuous transition." When I told Gregory Bateson what the Dalai Lama had said, his eyes widened in amazement and he said: "He's got it!"

So if we take a good look around us, we can observe the return of catastrophism to artistic and scientific narratives. What this means, I think, is that the rock bottom foundation for industrial society is giving out, and from the mathematics of René Thom to the novels of Doris Lessing we are being shown a new vision of planetary dynamics, a vision of sudden discontinuities. It is probably no accident that Ronald Reagan has come in to invoke all the old shibboleths of the industrial mentality, precisely at the moment when they are becoming inadequate, for one often sees in history that a radical shift is often preceded by an intensification of the old. Consider warfare in the fifteenth century: right at the moment when armor becomes most elaborate, with the knight lifted on to his horse by levers and pulleys, is right at the moment when the heavily armored knight is made irrelevant through the longbow, the crossbow, and firearms. Elsewhere I have called this kind of historical phenomenon "a sunset effect," but one can see it as a kind of supernova, an intensification of a phenomenon that does not lead to its continuation, but to its vanishing. So much for Reagan, but what about us? One theme that I think this conference could consider is the ways in which the new paradigm in science and art will relate to a new paradigm in politics.

But to return to the topic of narratives, let us consider a more humorous example than Darwin. Here is the syllabus for Professor William Buckland's lectures on geology at Oxford. This is how he outlines his course:

> First, the indications of the power, wisdom, and goodness of the Divinity would be demonstrated from the evidence of design in His works, and, particularly from the happy dispensation of coal, iron, and limestone, by which the Omnipotent Architect or Divine Engineer has assured manufacturing primacy to his British creations.[3]

That is Oxford University speaking, that is "the best and the brightest" of their time. I hope in the next century our culture will have grown to the point that the pronouncements of E. O. Wilson at Harvard will appear to be equally ludicrous.

I have called this opening talk "The Cultural Implications of the New Biology," but of course that is an oversimplification, for there is not simply one new biology, but several competing ones. But since we humans have a left and a right side, we tend to organize our world and our politics into sides. And so let me draw a rough sketch, a caricature if you will, of the new biology of the Right, and the new biology of the Left. On the Right is the sociobiology of E. O. Wilson; on the Left, the biologists gathered in this room. To appreciate the difference, let us contrast some statements by Wilson and some by Humberto Maturana. Here is Wilson:

> The transition from purely phenomenological to fundamental theory in sociology must await a full, neuronal explanation of the human brain. Only when the machinery can be torn down on paper at the level of the cell and put together again, will the properties of emotion and ethical judgment come clear.[4]

Notice how in Wilson's choice of language, reality can

only be perceived through an act of tearing down machinery to get at the basics. Reductionism is taken for granted as the primary act of understanding, and aggression is seen as the appropriate response of culture to nature. The archaic industrial mentality could not be more present, and so I see both Wilson and Reagan as part of the apologetics for the managerial system that joins science to state capitalism. Now let us take a look at Maturana:

> Neurons are the anatomical units of the nervous system, but are not the structural elements of its functioning. The structural elements of the functioning nervous system have not yet been defined, and it will probably be apparent when they are defined that they must be expressed in terms of invariants of relative activities between neurons, in some manner embodied in invariants of relations of interconnections, and not in terms of separate anatomical entities. In man-made systems this conceptual difficulty has not been so apparent because the system of relations (the theory) that integrates the parts that the describer (the observer) defines is provided by him, and is specified in his domain of interactions; as a consequence, these relations appear so obvious to the observer that he treats them as arising from the observation of the parts, and deludes himself, denying that he provides the unformulated theory that embodies the structure of the system which he projects onto them. In a self-referring system like a living system the situation is different: the observer can only make a description of his interactions with parts that he defines through interactions, but these parts lie in his cognitive domain only. Unless he explicitly or implicitly provides a theory that embodies the relational

> structure of the system, and conceptually supersedes his *description* of the components, he can never understand it. Accordingly, the full explanation of the organization of the nervous system (and of the organism) will not arise from any particular observation or detailed description and enumeration of its parts, but rather like any explanation, from the synthesis, conceptual or concrete, of a system that does what the nervous system (or the organism) does.[5]

As we can see, Wilson and Maturana are completely reversed images of one another, but contained in these two different biologies are two different ideas of procedure, two different ideas of order, and implicitly two different ideas of political order. Sociobiology denies the ontological value of the individual; all value rests in the gene pool and in the relationship of "inclusive fitness." The individual is simply a package for "the selfish gene." This view of the organization of parts to the whole is the worldview of a technocratic society, just as Darwin's perception of the struggle for survival was an expression of the worldview of an industrial society. Sociobiology is a form of apologetics for technocratic management: since the individual scientist himself cannot contain all the information of science, then Science becomes more important than the scientist. In this illusory world Science undergoes an apotheosis that raises it above the individual creative mind of the scientist, and the Scientific Method is canonized into a sanctified procedure that has very little to do with the actual way individual human scientists make discoveries and invent new theories.[6]

The role of unities is thus doubly denied by sociobiology: first, unities are broken down and defined in terms of fragments in the reductionist method, and, second, abstractions like species, gene pool, and inclusive fitness are reified and not simply seen as descriptive procedures of the observer. This world of broken fragments and chimerical abstractions is the frightening world of Science, capitalist or socialist, a world completely set apart from the organic processes of life in an ecology.

The biology of Maturana and Varela, by contrast, begins with the fundamental notion of unities. In Varela's *Principles of Biological Autonomy*,[7] he discusses the failure of conventional biology to recognize that the individual is the true ontological unity in evolution. We do not observe a species; we construct the idea of a species in an imaginary historical space. The creation of a species in a biological description is a performance of natural history that is related to cultural history. Here the individual takes on a new ontological value. In many ways there is an attractive link between the autopoesis of Maturana and Varela and the quantum theory of Heisenberg, for both share a more highly sophisticated epistemology. Heisenberg noted that we do not have any such thing as a "science of nature;" rather, we have a science of man's knowledge about nature. We do not live in reality, we live in a series of descriptions of reality. In their laboratory work concerning the physiology of perception, Maturana and Varela have provided us with some living examples of the way in which "reality" is a construct. The organism is much more than a Lockean template bombarded by so many outside meteor-like impressions. In one example they have shown that in color vision, the so-called signals come in at different times, and that the organism puts them together. Color vision is literally a synchronic construction.

Now, if we do not have a science of nature, but rather a science of humanity's knowledge about nature, then Science is not an external Jahweh-like deity that rules over us in an authoritarian fashion; it is a human activity, so human in fact, that we can more accurately say that natural history is a subset of cultural history, and not the other way round. The concept of "unity" becomes, therefore, an important perception that nature is made of processes rather than objects, and that these relational processes are always events within a domain of description by an observer. The sociobiologist looks for hard, irreducible objects that he can manipulate, but Heisenberg has said that the universe is made out of music, not matter; and so when a biologist sees processes and participates with them in a performance of descriptions, consciousness is participating with cognition in what Maturana and Varela would call "the realization of the living," or what Gregory Bateson liked to call Mind. And going back even farther beyond Gregory's work, we can relate these ideas of cognitive biology to the philosophy of organism in the work of A. N. Whitehead.

Color vision seems to be a singularly happy and painterly place to do research on the nature of process. Another one seems to be symbiosis. Traditionally, the majority of biologists have not liked the idea that the eukaryotic cell evolved through a process of organelles becoming endosymbionts, for it went against the grain in moving in the other direction of reductionism to atomistic parts. The bias toward objects and the blindness toward process made symbiosis a particularly difficult notion. The idea also was a direct affront to Social Darwinism, for it seemed to belong more to the point of view of Kropotkin in *Mutual Aid.* Sentences in Lynn Margulis's writing like: "Food scarcity in nature probably selects for the symbionts over the separate partners" is not in harmony with the value systems of an industrial society. This notion of food sharing is really fundamental to our biology and our politics. There is no more telling description than our idea of the origins of humanity, for how one sees the origins of human culture is also a description of how one wishes to see the future of humanity.

In anthropology there are two radically different conceptions of the origins of human culture. One is the view popularized by Robert Ardrey that it is the tool that made us human and separated culture from nature. In this view it is in the act of killing that we most truly perform our humanity. The weapon has a force of its own and it hurls its user into a new ecological niche, a new adaption, and all that is left behind is the sloughed-off animal nature of the primitive. The stone tool, exactly as was pictured in Kubrick's film *2001: A Space Odyssey*, is like a rocket which as it blasts toward heaven creates all hell below for those who happen to be underneath it. It is really only a tiny step from the anthropology of Ardrey to the triage of Garrett Hardin. Hardin has recently ridiculed the sentimental do-goodery of those who send food to Africa or Southeast Asia. For Hardin, this self-congratulatory act of the giver only increases the number of the suffering and prolongs their agony; it would be more charitable to drop atom bombs on them, for then their suffering would be over, and their numbers would be reduced to the proper carrying-capacity of the planet. Once again, we see the notion that in killing we truly express our humanity; a new atomic technology hurls a new elite into a new evolutionary adaptation, and once again the primitive apes are shoved aside to make room for the new technological culture. If we join the anthropology of Ardrey to the sociology of Hardin and the sociobiology of Wilson, we come up with these "stars" of thought joined in a constellation for a new historical landscape. We enter an epoch in which an authoritarian state is controlling world resources under the direction of the bureaucratic management of corporate science. From genetic engineering of populations of crops and peoples to triage, or from energy and power centralized in nuclear stations of troops and reactors, we get a clear picture of the crisis management of the world system in the last of the twentieth century.

But there is another view of the origins of human culture, and it too is a view that is connected to another vision of the future of humanity. Glynn Isaac in his essay on the food-sharing behavior of the proto-hominids,[8] has suggested that from his archaeological research in Africa there are indications that food was transported from one place to another where it was shared in circumstances of relative safety. Now here the basic act which makes us human is the sharing of food; small wonder that the religious among us feel that we most truly perform our human nature in the communion that comes from the sharing of food, whether it be the Jewish *seder* or the Christian eucharist. Interestingly enough in this week's issue of *Nature*, Henry Bunn has followed up on Isaac's work and says: "The documentation of meat-eating and the concentration of bones at particular places by early hominids lends strong support to the food-sharing model proposed by Isaac."[9]

In the technological definition of human culture, the tool basically separates culture from nature. In the social definition of human culture, the act of food-sharing is a relationship between nature and nurture. This vision of relationship is symbolically emphasized in the Christian sacrament, for when one takes raw grain and turns it into bread, one is involved in a performance of the movement from nature to culture; and so it is with the movement of the grape into wine. When Jesus takes bread and wine and says: "Take this in remembrance of me, for this is my body and blood," he is not the masochistic psychopath that Freud made him out to be, but a poet with an ecological vision of life who is using myth and symbol to express how all life is food to one another. The *Upanishads* would express the idea in a different poetic diction to say that: "Earth is food; air lives on earth; earth is air, air is earth; they are food for one another." The theological idea, then, of the Mystical Body of Christ put forth by St. Paul, is a vision of a planetary being, a cell in which we as individuals are organelles.

Now, if food-sharing is the fountainhead and source of our original humanity, then we most truly perform that humanity when we share food and see with Lewis Thomas in his *Lives of a Cell* that the whole Earth is a single cell and that we all are simply symbiotic organelles involved with one another. There can be no "us" and "them." The global politics that issues forth from this vision is truly a *bios* and a *logos*.

It is, therefore, small wonder that many scientists have rejected the theory of "symbiosis and cell evolution," and that Lynn Margulis has had to work hard for over a decade to win support in favor of her views.[10] The rejection of symbiosis and the rejection of autopoesis are both expressions of a single mind-set: the preferential perception of objects over processes, of fragments over constitutive relationships, of technology and control over epistemology and understanding.

So, as we can see, the two biologies are implicitly two different politics because they are essentially two different worldviews. One provides the scientific apologetics for the crisis management of the disintegrating modern world system; the other provides the scientific foundation for the politics of a new planetary culture. One is emphasized and strongly supported by the universities and governments; the other is emphasized by Lindisfarne, a group that, as we all know, lacks the funds to support any scientific work.

Both technocratic state capitalism and scientific socialism are locked into an industrial worldview. The Marxists in Europe have tried to absorb the ecological issue into their political rhetoric, thinking that this could provide them with a new constituency and a broader popular base, but ecology represents a different way of thinking, and in the case of the biology of knowledge, a radically different way of regarding structure and organization. These new ideas simply do not fit in with dialectical materialism and scientific socialism; the European intelligentsia of the Left is simply going to have to exercise its celebrated intelligence to rethink the whole matter of mind along the lines suggested by Gregory Bateson in the last section of *Steps to an Ecology of Mind*. The genetic engineering of crops and the sociobiological management of society can fit in very nicely with a new form of Marxist-Leninism, and certainly the crisis management of our American society may reach the point of cultural merger through technological procedures, but the Gaia hypothesis of Lovelock and Margulis, the cognitive biology of Maturana and Varela, and the ecology of Jackson and Todd lead more in the direction of the gentle anarchism of Kropotkin than to the scientific socialism of Trotsky and Lenin. So, in this sense, Lindisfarne is neither Left nor Right, reactionary nor avant-garde, but all of those at once.

The fundamental principle that I see coming out of this new mode of thought is that living systems express a dynamic in which opposites are basic and opposition essential. One cannot say that the ocean is right and the continent is wrong in a Gaian view of planetary process. What this means for me is that the movement from archaic industrial modes of thought into a new planetary culture is characterized by a movement from ideology to an ecology of consciousness. In ideological modes of thought one believes that the Truth can be expressed in an ideology, and that that ideology can be administered to the masses by an elite that is pure and true to that ideology. It doesn't matter whether we are talking about ayatollahs in Iran, communists in the Politburo, capitalists in the Hoover Institute, or terrorists in the Red Guard; the structure of thought is the same, only the content has been changed. Now in a polity that is formed by the shapes of interacting opposites, an *enantio-morphic* polity, the fundamental notion is that the Truth cannot be expressed except in relationships of opposites; therefore, every ideology is partial and in its purest elaboration it is most incomplete. As Niels Bohr expressed the idea a generation ago: "The opposite of a fact is a falsehood, but the opposite of one profound truth may well be another profound truth." Capitalism and communism, or Judaism and Islam, are both simultaneously right, and any new world system is going to have to become an ecology of consciousness in which these opposites interact in non-annihilating ways.

The second principle of an enantiomorphic polity is expressed in the principle of hierarchy. Hierarchy is a structural system that takes energy from a dangerous or unusable level, steps it down and makes it available for work at a lower and more generalized level. For example, the earth's atmosphere takes the sun's energy, steps it down and makes it available for a life without skin cancer. The necessary opposite to this hierarchical principle is the holographic one: every microcosm mirrors the macrocosm; the initiate or genius may exist as a transformer, but Godhead is equally present in all. When therefore political hierarchies try to reify spiritual or cultural ones in a pharaohonic way, there is released an equal and opposite energy which could be called "redemption through the primitive." In history we have seen this as a pattern of tribes against the empire. In the holographic principle there is a recursiveness of energy; if the energy is perverted into a simple top-down pyramid, then the system breaks apart in revolution. The icon for this relationship of opposites is not the pyramid, but the double triangles of Yeats's *A Vision.*

The third principle of an enantiomorphic polity would not be the vertical one of transcendence, but the horizontal one of immanence: values are not objects but moiré patterns which emerge from the superimposition of opposites. When capitalist and communist fight with one another, the truth is in neither one, but in the moiré pattern they establish in their *agon.* The truth is in the system, not out of it, and both the individual and the collective are needed to express the process we call life.

The fourth principle is a temporal one, the idea of the enantiodromia that Jung reintroduced from alchemy and that I discussed in *Evil and World Order.*[11] This idea simply observes that every social process for the achievement of value, when fully developed, turns into its opposite. The Ayatollah Khomeini becomes another Shah.

Had the Ayatollah not tried to reify the principle of hierarchy in a theocratic state, had he seen the need for the double triangles (which, ironically, are present in the Shi'ite philosophy as explicated by Henry Corbin), he would not have so remorselessly turned into his enemy.

In order for the grammar of the English language to inform my sentences, it must stay out of them; and so it is with the spiritual grammar of Being; it can only inspire politics through the fullness of culture; when spiritual principles are reified into a political hierarchy, one gets a horrible perversion in the confusion of levels. One gets contemporary Iran, or the Catholic Church of the Inquisition, or the Aztec state. Since we become what we hate and the force of our own passion turns us into our own enemy, then the dynamics of the enantiodromia mechanically keep the wheel of *samsara* turning as Ayatollah becomes Shah. The call to a politics of enlightenment is the call of Jesus to love one's enemy, to move from mindless passionate conflict to mindful dispassionate opposition, from eros to agapé. William Blake understood Jesus better than most when he said: "In opposition is true friendship" or "Without contraries is no progression."

These principles for an enantiomorphic polity are not philosophical rules to be enforced by new revolutionary guards, but simply perspectives of the political compassion necessary for a new healthy world culture. At the end of his life, our former Lindisfarne Fellow, E. F. Schumacher, came to a similar point of view.

> The pairs of opposites, of which *freedom and order* and *growth and decay* are the most basic, put tension into the world, a tension that sharpens man's sensitivity and increases his self-awareness. No real understanding is possible without awareness of these pairs of opposites which permeate everything man does.
>
> In the life of societies there is the need for

> both justice and mercy. "Justice without mercy," said Thomas Aquinas, "is cruelty; mercy without justice is the mother of dissolution"—a very clear identification of a divergent problem. Justice is a denial of mercy, and mercy is a denial of justice. Only a higher force can reconcile these opposites: wisdom. The problem cannot be solved, but wisdom can transcend it. Similarly, societies need stability *and* change, tradition *and* innovation, public interest *and* private interest, planning *and* laissez-faire, order *and* freedom, growth *and* decay. Everywhere society's health depends on the simultaneous pursuit of mutually opposed activities or aims. The adoption of a final solution means a kind of death sentence for man's humanity and spells either cruelty or dissolution, generally both.[12]

And now let me place Schumacher's ideas alongside Henri Atlan's synthesis of information theory and biology. This is from Atlan's *Entre le cristal et la fumée,* but I'll present it in English.

> Thus it suffices to consider organization as an uninterrupted process of disorganization, organization, and not as a state. For order and disorder, the organized and the contingent, construction and destruction, life and death, are not that distinct anymore; and moreover that is not all of it. These processes in which is realized the unity of opposites—these do not become a new state, a synthesis of the thesis and the antithesis. These processes cannot exist, except that errors are *a priori* true errors, that order at any given moment is truly disturbed by disorder, that destruction, albeit not totally realized, is still real, that the eruption of the

> event is a true eruption, a catastrophe, or a miracle, or both. In other words, these processes which appear to us as the foundations of the organizations of living beings, resulting from a sort of collaboration between what one has the habit of calling life and death, can only exist if it's never a question simply of collaboration, but always one of radical opposition and negation.[13]

Here we are not presented with some nice and comforting liberal compromise, we are presented with a more Greek tragic vision, one in which the Truth is expressed only to the degree that the polarities are truly polarized with enough energy to allow the magnetic field of the whole to spread itself before us. Atlan's is a clear vision of dialectic and negation. Nevertheless, there are limits; there are parameters of restraint that keep living systems as living systems. If we move from Blake's moony Beulah where all contraries are true, because the soul is at rest, to the vibrancy of life in a New Age, then we move from Blake's corporeal to mental war in an alchemical transformation.

Through Spirit the world has always been one, and now through electronic technology the world has learned again to look at itself as one. But we do not yet have a politics in keeping with our spirituality, our art, our science, or our technology. And this seems to be the work that is cut out for our generation:

> Rouze up, O Young Men of the New Age! Set your foreheads against the ignorant Hirelings! For we have Hirelings in the Camp, the Court & the University, who would, if they could, for ever depress Mental & prolong Corporeal War.[14]

To move from a condition of conflicting ideologies to an ecology of consciousness on a global level is going to require an enlightenment more profound than the philosophical Enlightenment in Europe that inspired the American and French Revolutions. Some progress is being made in this direction, but we are still in the early days of the shift from one world system to another. Slowly the world is organizing itself into transnational blocs. At the moment these blocs are economies that are creating forms of association beyond regional cultures; they are not yet planetary cultures within a global ecology of consciousness. Nevertheless, a fourfold pattern of grouping does seem to be apparent.

1 The Capitalist World (U.S.A., Western Europe, Japan, the countries of the Pacific Rim, Korea, Taiwan, Australia).
2 The Communist World (U.S.S.R., Eastern Europe, China, parts of Africa and Latin America).
3 The Resource-Rich World.
4 The Resource-Poor World, or the "Least Developed Countries."

In the third bloc both economies and religion form patterns of transnational association, and just how antimodernizing Islam and industrial oil will trade off with one another is going to be a challenge to the ideologies of both capitalism and socialism. But, in any event, the world is too vast and complex to be dominated singly by any one of these four blocs, whatever the megalomaniacal dreams may be of a universal Islam, a universal communism, or a universal fundamentalist Christianity.

In the transition from ideology to an ecology of consciousness, we shall probably see a shadow-form that will, most likely, be the rule of Technique. Here the technician will counsel us to put aside our conflicting ideologies to move up to the more advanced world of technology. The rule of Technique is, as Jacques Ellul and Ivan Illich have shown, so much camouflage for the rule of a particular managerial class, and in this monocrop mentality one ideology is simply replaced with another one. This is why the global mechanists, the Bucky Fullers and the Jay Foresters, cannot take us all the way into the new planetary culture. To move into an ecology of consciousness we need also the suggestions of the planetary mystics. To understand non-ideology in the polity, we must understand non-ego in the person, and that is where Buddhism takes on a new relevance to forms of education in an electronic culture beyond the Reformation with its "Protestant Ethic and the Spirit of Capitalism." This is the reason why in Lindisfarne's work in education we have worked, not simply for a dialogue among the world's religions, for that is too bureaucratic, but for a contemplative association. And that is also why in the Lindisfarne Fellowship we have worked to create, not a new ideology to which we all subscribe, but an ecology of associated differences. And so we have this meeting on the new biology in the middle of a Zen Buddhist monastery.

A long time ago, in the shift from the medieval to the modern world system, the small school of Ficino's Academy in Renaissance Florence served to gather poets and philosophers to envision a new culture. A few centuries later, a group of thinkers with Franklin and Jefferson came together in the American Philosophical Society to envision a new democratic society. Now as we move into the period of crisis for the modern world system of industrial nation-states, a period not simply of wars of resources, but also of ecological planetary damage from unbalanced industrialization, we will need to come together to envision a new world. What physics was to engineering in industrial society, biology has become to ecology in our new society. As we move from economics to ecology as the governing science of our era of stewardship, our politics will have to help us realize, beyond all budgets and bottom lines, that what truly counts can't be counted.

Notes

1. As quoted in *Darwin: The Norton Critical Anthology* (New York, Norton, 1970), p. 169.
2. Ibid., p. 198.
3. As quoted in C. C. Gillispie's *Genesis and Geology* (New York, Harper & Row, 1959), p. 104.
4. E. O. Wilson, *Sociobiology: The New Synthesis* (Cambridge, Harvard, 1975), p. 575.
5. Humberto Maturana and Francisco Varela, *Autopoesis and Cognition: The Realization of the Living* (Dordrecht, Holland, Reidl & Co., 1980), p. 47.
6. See Paul Feyerabend, *Against Method* (London, Verso Edition, 1978).
7. Francisco Varela, *Principles of Biological Autonomy* (New York, Elsevier-Holland, 1979), p. 39.
8. Glynn Isaac, "The Food-Sharing Behavior of Proto-Hominids," *Scientific American,* April, 1978, Vol. 238, No. 4, pp. 90–108.
9. Henry T. Bunn, "Archaeological Evidence for Meat-Eating by Plio-Pleistocene Hominids from Koobi Fora and Olduvai Gorge," *Nature,* Vol. 291, June 18, 1981, p. 576.
10. Lynn Margulis, *Symbiosis and Cell Evolution* (San Francisco, Freeman, 1981).
11. William Irwin Thompson, *Evil and World Order* (New York, Harper & Row, 1976), p. 79.
12. E. F. Schumacher, *A Guide for the Perplexed* (New York, Harper & Row, 1977), p. 127.
13. Henri Atlan, *Entre le cristal et la fumée* (Paris, Edition de Seuil, 1979), p. 57 f.
14. William Blake, Preface to *Milton* in *The Poetry and Prose of William Blake,* ed. David Erdman (New York, Doubleday, 1965), p. 94.

PART ONE

Biology and the Way of Knowing

1

GREGORY BATESON

Men Are Grass

Metaphor and the World of Mental Process

THIS IS A tape of a lecture intended to be given at the Lindisfarne Fellows' Meeting at Green Gulch in June 1980. I wish I were there with you, but when it appeared that I very likely would not be able to get to Green Gulch for this meeting, I talked to Bill Thompson and suggested that I dictate a tape to be played if he should so desire—failing which, I am sure that somebody else in this room is very capable of taking the first pitch in talking to you at this meeting. Bill advised me that I should talk about what has most been exercising my mind in the last two or three months, and offer that to you as a basis for your discussions. I have had two things on my mind. One is very general, perhaps too general, and the other rather specific. If I were there among you, I would prefer to speak mostly on the specific matter, hoping for discussion which I could use, but since that is apparently not to be, let me offer you the general matter, which, in effect, becomes a survey of almost everything I've done in my life. A survey of a direction in which I have tried always to be moving, though that direction, of course, gets redefined and redefined from project to project.

I grew up in the middle of Mendelian genetics. And the vocabulary that we used then was a curious one. We used to speak of Mendelian factors. Now the word "factor" was a word coined to avoid saying "cause," and at the same time to avoid saying "idea" or "command." You will remember that in the nineteenth century there had been deep and bloodthirsty battles around the Lamarckian concept of the inheritance of acquired characteristics. And this concept was tabooed because it was believed, I think incorrectly, that it necessarily introduced a supernatural component into biological explanation. This component was variously called "memory," "mind," and so forth, but I don't believe it was a supernatural component. It would seem to me to fit with very little modification into the general scene of biological explanation, though its fitting would indeed alter the basis of biology from the very ground up and would alter our ideas about our relationship to mind, our relationship to each other, our relationship to free will, and so on. In a word, our complete epistemology. Here in what I have just said you will notice the assumption that epistemology and theories of mind and theories of evolution are very close to being the same thing, and epistemology is a somewhat more general term which will cover both the theories of evolution and the theories of mind.

The battles over this battleground had been fierce and bloody, and, with a few exceptions, nobody wanted to go through those battles again. So we are still going through them. In any case, it seemed safer at that time to refer to the causal agencies, or the explanatory components of genetics, as factors rather than commands or memories. Darwin, as you may know, had funked the issue of mind and matter in the last pages of the *Origin of Species*. There he suggests that while his evolutionary theory accounted for what had happened to living things once biological evolution had started on the face of the earth, it is possible that that vast inheritance did not start on earth, but reached earth in the form of bacteria, perhaps riding on light waves or whatever, a theory which I've always felt was a little childish. I've been told by a member of the Darwin family that it was probably put in because he was afraid of his wife, who was an ardent Christian. Be all that as it may, the mind/body problem or the mind/matter problem was avoided in those early days of the twentieth century. It is still largely avoided in zoological schools, and the terms "Mendelian factor," "allele," etc., were all rather convenient euphemisms to avoid acknowledging that the field of inquiry was split wide open.

My father, in the 1890s, had set out to do approximately (and this is really very strange) what I have been trying to do in the last few months. Namely, to ask, if we separate off, for the sake of inquiry, the world of mental process from the world of cause and matter, what will that world of mental process look like? And he would have called it, I think, the laws of biological variation, and I would be willing to accept that title for what I am doing, including, perhaps, both biological and mental variation, lest we ever forget that thinking is mental variation.

And, of course, I walk into this field with a lot of tools that my father never had. It perhaps is worth listing these quickly: there's the whole of cybernetics, there's the whole of information theory, and that related field which I suppose we might call communication theory, though as you will see I don't much like the word. Organization theory would be a little better, resonance theory perhaps a little better still. In addition, and very importantly, I have a rather different attitude towards Lamarck, and towards the supernatural, and towards "God." A hundred years ago these things were dangerous to think about, and there was a feeling that how one classified them could be wrong. Personally, I feel that how one classifies the inheritance of acquired characteristics (is it a case of ESP?) is largely a matter of taste, but has tied into its tail, like all matters of taste, the threat that there are many ways of performing this classification which will in fact lead to disaster. If you want to call these ways of classifying wrong, it's all right by me, but personally I want to know more about the total mental web we're talking about, so that the word "wrong" or the words "bad taste" or whatever shall take on a natural history meaning. And that's what I'm really trying to do, to discover, to explore. And I start from a position which is a little more free to take an overall view than was the position of the previous generation.

Then too, I start from the position in which I have some idea of the nature of what I want to call "information." Namely, that this "stuff" is precisely not *that*, a thing, and that the entire language of materialism, good as it may be for the description of relations between material things, reflecting back upon the things, is lousy as a way of describing the relations between things to reflect forward upon their organization. In other words, the entire materialistic or mechanistic language is inappropriate for my use, and I simply have to have the courage to discard it. This means, of course, that in my mental world or universe I acknowledge no things, and, obviously, of course, there are no things in thought. The neurons may be channels for something, but they are not themselves things within the domain of thought, unless you think about them, which is another thing again. In thought what we have are ideas. There are no pigs, no coconut palms, no people, no books, no pins, no . . . you know? Nothing. There are only ideas of pigs and coconut palms and people and whatever. Only ideas, names, and things like that. This lands you in a world which is totally strange. I find myself running screaming from its contemplation, and essentially running back to a world of materialism, which seems to be what everybody else does, limited only by the amount of their discipline. What I feel driven to ask is, give me a pound, a little mass, a little time, a little length, some combination of these called energy. Give me power, give me all the rest of it. Give me location, for in the mental world there is no location. There is only yes and no, only ideas of ideas, only news of messages; and the news is news, essentially, of differences, or difference between differences, and so on. What is perpetually happening in the works of the most learned philosophers, as well as of people like myself, is a quick dash back into the idioms and styles and concepts of mechanical materialism to escape from the incredible bareness—at first appearance—of the mental world.

Now, notice that in this throwing out of our favorite devices for explanation, a lot of very familiar stuff upon which we are deeply dependent has gone out with the bath water, and I think good riddance. Notably the separation between God and His creation: that sort of thing doesn't exist anymore. Notably the separation between mind and matter: we won't be bothered by that anymore except to look at it with curiosity as a monstrous idea that nearly killed us. And so on.

I think it is time that I provided my mental world with a little furniture. So far all you've had is the idea that it is full of ideas and messages and news, and that the intangible filter which is between the material and mechanical world and the world of mental process is simply this filter of difference. That while ten pounds of oats is in the sense of materialism real, the ratio (and I repeat the word ratio: I don't mean the subtractive difference—the contrast, if you like, yes) between five pounds and ten pounds is not a part of the material world. It does not have mass, it does not have any other physical characteristics—it is an idea. And there is always this shift to a first derivative between the mechanical world and the world of mental process. I derived this point in about

1970 from Alfred Korzybski. Those of you who are here may remember the Lindisfarne meeting where A. M. Young and I had a confrontation, I think a rather unfortunate one perhaps. He was saying very much the same thing as this and extending it in certain ways which meant that he was going to, as I saw it, forget the rule of dimensions, and indeed the whole of logical typing, in his understanding of mental life. I thought that was a very severe mistake: I don't know who was right. In any case, that's the first positive characteristic that I have given you about the mental world.

Let me now bring in another whole family of descriptive propositions, descriptive of epistemology, about which it's not quite clear whether they belong on the mechanical side or on the side of mental process. I favor the latter, but let's consider it. These are the propositions which St. Augustine, a very long time ago, called Eternal Verities, of which Warren McCulloch, a dear friend of mine, was always fond, if you can be fond of anything quite so impersonal. The Eternal Verities of St. Augustine were such propositions as "three and seven are ten." And he averred that they had always been ten and always would be ten. He was not interested, of course, in this division between the mental and the mechanical or physical that I am talking about, so he didn't touch that as far as I know. But we are interested in that. My feeling is that there is a contrast between what I call quantity and what I call pattern, and in this contrast I see number, at least in its simplest forms, smaller forms, as inevitably of the category and nature of pattern rather than of the nature of quantity. So number is perhaps the simplest of all patterns. In any case, St. Augustine was a mathematician, and in particular an arithmetician, and he seems to have had a feeling that the numbers were very special things, a feeling, of course, which is not unfamiliar to most of you who have thought a little about Pythagorean numerology and other related things. Then, after all, the contrasts between numbers are very much more complex than the mere ratios. We could say, I suppose, that the contrasts—pattern differences—between numbers fall off as the numbers get bigger and bigger, but I'm not sure the numerologists will permit us to say that. What seems to be clear is that at least in smaller numbers the pattern differences, between say three and five, are drastic indeed, and form in fact major taxonomic criteria in biological fields. I am after all interested in this realm of pattern or number or mental process as a biological realm, and the biological creatures, plants and animals, certainly seem to think that their concern is much more with number than with quantity, though above a certain quantitative level, a certain size of number, as I pointed out in *Mind and Nature*, numbers become quantities, so that a rose has five sepals, five petals, many stamens, and then a gynoecium of a pistil system based on five. The contrast between four sides of a square and three sides of a triangle is not four minus three, being one, it is not even the ratio between four and three. It is very elaborate differences of pattern and symmetry which obtain between the two numbers as patterns.

So it would seem that this pattern aspect of numbers at least belongs in the mental world of organisms. Now I want to introduce into that world another component, which I confess is rather surprising. It's been clear for a long time that logic was a most elegant tool for the description of lineal systems of causation—if A, then B, of if A and B, then C, and so on. That logic could be used for the description of biological pattern and biological event has never been at all clear. Indeed it is rather sharply clear that it is unsuitable, at least in the descripton of such circular causal systems and recursive systems as will generate the paradoxes. Now, for those you can muddle along, maybe completely, I don't know, with a correction of the lineal system by appeal to time. You can conclude your Epimenides paradox with the statement: yes at time A, and if yes at time A, then no at time B; if no at time B, then yes at time C; and so on. But I do not believe that this is really how it's done in nature. I mean you can do it on any page of your book, but it's another thing to say that these are the logical causal trains, or whatever, which in fact occur in organisms and their relationships and their tautologies of embryology and so on. You will see that this is a very unlikely solution.

On the other hand there is another solution which I would like to present to you. Would somebody please place on the blackboard these two syllogisms side by side. The first is a syllogism in the mood traditionally called Barbara:

Men die.
Socrates is a man.
Socrates will die.

And the other syllogism has, I believe, a rather disreputable name, which I will discuss in a minute, and it goes like this:

Grass dies.
Men die.
Men are grass.

Thank you. Now, these two syllogisms coexist in an uncomfortable world, and a reviewer the other day in England pointed out to me that most of my thinking takes the form of the second kind of sequence and that this would be all very well if I were a poet, but is inelegant in a biologist. Now, it is true that the schoolmen or somebody took a look at various sorts of syllogisms, whose names are now, thank God, forgotten, and they pointed to the "syllogism in grass," as I will call this mood, and said, "That's bad, that does not hold water, it's not sound for use in proofs. It isn't sound logic." And my reviewer said that this is the way that Gregory Bateson likes to think and we are unconvinced.

Well, I had to agree that this is the way I think, and I wasn't quite sure what he meant by the word "convinced." That, perhaps, is a characteristic of logic, but not of all forms of thought. So I took a very good look at this second type of syllogism, which is called, incidentally, "Affirming the Consequent." And it seemed to me that indeed this was the way I did much of my thinking, and it also seemed to me to be the way the poets did their thinking. It also seemed to me to have another name, and its name was metaphor. Meta-phor. And it seemed that perhaps, while not always logically sound, it might be a very useful contribution to the principles of life. Life, perhaps, doesn't always ask what is logically sound. I'd be very surprised if it did.

Now, with these questions in mind, I began to just sort of look around. Let me say that the syllogism in grass has a quite interesting history. It was really picked up by a man named E. von Domarus, a Dutch psychiatrist in the first half of this century who wrote an essay in a very interesting little book, which has more or less disappeared, called *Language and Thought in Schizophrenia.*[1] And what he pointed out was that schizophrenics tend, indeed, to talk, perhaps also to think, in syllogisms having the general structure of the syllogism in grass. And he took a good look at the structure of this syllogism, and he found that it differs from the Socrates syllogism, in that the Socrates syllogism identifies Socrates as a member of a class, and neatly places him in the class of those who will die, whereas the grass syllogism is not really concerned with classification in the same way. The grass syllogism is concerned with the equation of predicates, not of classes and subjects of sentences, but with the identification of predicates. Dies—dies, that which dies is equal to that other thing which dies. And von Domarus, being a nice, you know, honest man, said this is very bad, and it is the way poets think, it's the way schizophrenics think, and we should avoid it. Perhaps.

You see, if it be so that the grass syllogism does not

require subjects as the stuff of its building, and if it be so that the Barbara syllogism (the Socrates syllogism) does require subjects, then it will also be so that the Barbara syllogism could never be much use in a biological world until the invention of language and the separation of subjects from predicates. In other words, it looks as though until 100,000 years ago, perhaps at most 1,000,000 years ago, there were no Barbara syllogisms in the world, and there were only Bateson's kind, and still the organisms got along all right. They managed to organize themselves in their embryology to have two eyes, one on each side of a nose. They managed to organize themselves in their evolution. So there were shared predicates between the horse and the man, which zoologists today call homology. And it became evident that metaphor was not just pretty poetry, it was not either good or bad logic, but was in fact the logic upon which the biological world had been built, the main characteristic and organizing glue of this world of mental process which I have been trying to sketch for you in some way or another.

Well, I hope that may have given you some entertainment, something to think about, and I hope it may have done something to set you free from thinking in material and logical terms, in the syntax and terminology of mechanics, when you are in fact trying to think about living things.

That's all.

Notes

This essay is the transcription of the tape made by Gregory Bateson as the opening address of the annual meeting of the Lindisfarne Fellows on June 9, 1980, at Wheelwright Center, Green Gulch. It was dictated a few weeks earlier at Esalen Institute in Big Sur when he realized that his health was not going to allow him to attend the gathering in person. Bateson died at noon on July 4, 1980, at the guest house of San Francisco Zen Center.

1. E. von Domarus, "The Specific Laws of Logic in Schizophrenia," in *Language and Thought in Schizophrenia,* ed. Jacob Kasanin (Los Angeles and Berkeley, University of California Press, 1944).

2

FRANCISCO VARELA

Laying Down a Path in Walking

The great sea
Has sent me adrift,
It moves me as the weed in a great river,
Earth and the great weather move me,
Have carried me away,
And move my inward parts with joy.[1]

LIKE A FUGUE we hear from afar, the transition from where we are to where we shall be is ruled by a few chords that play over and over again, everywhere.

What moves me in the poem I have chosen as an epigraph is the swift somersault between the so-called inner and outer, between mind and nature, between rocks and bowels. Where do we find here the proud distance between us and it? There is no distance, not even the distance between an it and its picture, which makes it possible to ask how accurate a representation the picture is. The theme of the fugue I am hearing, then, moves past a split Cartesianism to give flesh to a world of no-distance by mutual interdefinition.

In these pages I intend to spell out this theme as it plays in biology, and the way in which it shapes some fundamental problems. This is what I understand to be the "new biology." It is a ferment of the current dynamics of biological research. The introductory essay by W. I. Thompson was similar in intention, but I shall speak here as a research biologist, and not as a cultural historian.

Let me make a confession before I plunge into the subject. I am a bigot in epistemology. To me, the chance of surviving with dignity on this planet hinges on the acquisition of a new mind. This new mind must be wrought, among other things, from a radically different epistemology which will inform relevant actions. Thus, over and above their intrinsic beauty, I take these epistemological meanderings as vital. Literally. Therefore, in the discussion that follows, I would like you to direct your mind to the same place you would if the discussion were about, say, animistic cosmology. Our current notions about evolution and brain will be as distant to our grandchildren as this animistic cosmology is to us today.

My strategy for leading you in the direction I am looking will be as follows. First, I shall present a rough sketch of the main issues involved through the use of a metaphor disguised as a thought experiment. Second, I shall show how these issues take flesh in the current notions of evolution and its alterations, and, third, I shall examine the brain sciences from a similar perspective. The choice of both of these areas of biology is, of course, no accident, for evolution and cognition are really flip sides of the same conceptual coin (as Gregory Bateson was fond of reminding us). In the fourth and last section of the paper, you will have, I trust, new conceptual goggles, so that when we come back to the main issues again, they will be virtually redundant to you. You will be able to state them in your own language, for your own concerns.

A First Glimpse into Autonomous Unity

A simple, yet quite accurate way to state what I see as the pivot of the transition from the old (half a century) biology into a new one is as follows. Instead of being mainly concerned with heteronomous units which relate to their world by the logic of correspondence, the new biology is concerned with the autonomous units which operate by the logic of coherence. Thus the contrast I am proposing is:

CURRENT BIOLOGY	Heteronomous units operating by a logic of correspondence
NEW BIOLOGY	Autonomous units operating by a logic of coherence

Now, I might as well have written that in Martian, for those two aphoristic remarks are too densely packed. Let us move on to unfold the remarks by the aid of a thought experiment.[2]

Imagine in your mind's eye and ear a mobile, with thin pieces of glass dangling like leaves off branches, which dangle from other branches, and so on. Any gust of wind will cause the mobile to tinkle, the whole structure changing its position, speed, torsion of the branches, etc.

Clearly, how the mobile sounds is not determined or instructed by the wind or the gentle push we may give it. The way it sounds has more to do with (is easier to understand in terms of) the kinds of structural configurations it has when it receives a perturbation or imbalance. Every mobile will have a typical melody and tone proper to its constitution. In other words, it is obvious in this example that in order to understand the sound patterns we hear, we turn to the nature of the chimes, and not to the wind that hits them.

But let's carry this *Gedanken* experiment just one step further, and imagine now that the intricate structure of leaves and branches full of tinklers has the unusual capacity of moving the entire thing over the ceiling where it hangs. This could be accomplished, say, through detachable air-sucking devices which are alternately pressurized and depressurized. Thus, in this improved mobile-mobile, any gust of wind will produce not only a tinkling sound, but also a motion in some direction.

Wouldn't it be a surprise if we find that the whole mobile-mobile is moving with some sensible (to us) behavior? For instance, each time a wind blows, the mobile moves around until it finds a place with less wind, or, conversely, it searches for the origin of the air current and thus delights us with almost perpetual melodies.

If this mobile-mobile wind chime were to show such behavior, we would conclude that someone has designed it with cunning imagination so that it can do what it does. It seems utterly inconceivable that a mobile could come up with such smart motions by random arrangement of leaves, branches, and air-sucking devices.

The point of this example is to suggest the relative ease with which a degree of self-involvedness immediately gives the system a desire for autonomy vis-à-vis its medium. That is to say, the fact that it handles its medium according to its internal structure becomes the predominant phenomenon. If you think of the mobile-mobile as having a perception of the world, then clearly perception is not a matter of what gets into it, like, for example, an instruction for a man-made device. Perception has to do, rather, with how the system is put together, and, moreover, with how it perceives itself, in the sense that its own entanglement is the key to understanding what will happen to it.

A second point of the example is to realize that, should an apparently sensible behavior arise, the temptation is to say that it has been engineered in some way. Let us examine this point more closely by introducing the last complication into our thought experiment, as follows. I now assure you that in the case of this mobile-mobile which exhibits such an interesting behavior there has been no design whatsoever; indeed, the structural configuration exhibiting such interesting behavior patterns was arrived at by pure trial and error, a sort of tinkering with the shapes of the branches and the interconnections with the air-sucking devices. What are we then to say?

The traditional explanation (or description) of the situation would be that the system has some degree of internal representation of the physical environment, so that it knows how to respond to the wind. It has a correspondence to the world through a simple mirroring of some of its qualities. The mobile-mobile has become a representational system, that is to say: an active, self-updating collection of structures capable of "mirroring" the world as it changes. Now, if there had been an engineer who had actually figured out how to put the branches together so as to produce this behavior, such a description would seem appropriate. But, *ex-hypothesi*, the system came into being by mere tinkering, not design. How then are we to approach this situation?

We need a subtle but powerful twist: we emphasize the system's coherence, instead of taking the perspective of a supposed design. In other words, we understand the system as an autonomous cognitive system: an active, self-updating collection of structures capable of informing (or shaping) its surrounding medium into a world through a history of structural coupling with it.

These, then, are two alternative modes of description. One supposes mirroring and representation of features which are relevant and visible to us as observers, and requires, in some form or another, an agent which designs, because it requires a perspective from which this correspondence of world to the innards of the system is established *ex-professo*. The second perspective is more parsimonious. It states that out of the many possible paths of tinkering, the particular one we observe allows us to see what is a world for the system, that is, the particular way in which it has maintained a continuous history of coupling with its medium without disintegration. There is no mirroring, but in-forming. The first description hinges on a logic of correspondence; the other on a logic by coherence.

There is more than meets the eye in this *Gedanken* experiment. It really underscores a change in attitude and framework that has ramified implications, as we shall presently see. The reasons for this are simple: we have changed our point of view from an externally instructed unit with an independent environment linked to a privileged observer, to an autonomous unit with an environment whose features are inseparable from the history of coupling with that unit, and thus with no privileged perspective. In so doing, we are also on our way to spelling out a mechanism by which cognitive processes can be understood and built, a mechanism by which unities can endow a world with a sense through their structure and history of interactions.

A description by correspondence is essential for relating to units such as computers and washing machines (until they break down), but it turns out to be a rather limiting framework to use when it comes to life and mind (that is, for almost everything). Let us now turn to what this framework does for our understanding of evolution and the brain.

A Walk through the Adaptationist Program and Back Again

Think for a moment of the bar scene in *Star Wars*; picture the beings present there, and let us look at them through the eyes of a zoologist. The most obvious observation is that they are essentially of one kind: vertebrate-like. There are wild varieties in dermatological appearance—type of skin, shape of eyes—but they stand up straight, and most of them even look warm-blooded. How a culture conceives of imaginary beings is a clear indication of its conception of life, because it sets off the limits of what is imaginable. In seventeenth century zoology texts, next to eagles and chickens, we can find beings with human bodies bearing birds' heads. It was all conceivable, part of the same nature.[3] In the twentieth century showcase of imaginary zoology in *Star Wars*, we see nothing of the sort: uniformity is the essential guiding principle.

The point of this digression is to introduce the idea that in our culture at large—including science—we see ourselves as the best and only possible way of being intelligent. We have come up from modest beginnings through a direct path of optimization in an evolution guided by natural selection. What are the biological roots of this commonsensical understanding? The answer to that question lies in the main characteristics of evolutionary thinking over the last half century, which is in fact not difficult to state: the search for optimal mechanisms of adaptation to the world. Let me explain.

Stated bluntly, this approach assumes that species and communities have become, through their history, optimally adapted to their niche. The job of the evolutionist is to find the precise ways in which this process has occurred. It is not a matter of if, but how. Natural selection is seen as an ingenious engineer or smart gambler in the game of life versus environment (without assuming an external purpose, of course). The search for this optimization most commonly takes the form of isolating a specific trait from the organism's morphology, physiology, or behavior, and finding what it is optimum for and how. For example, one shows that the shape of cilia in protozoans is such that they are at their hydrodynamic optimum. (This sometimes gives rise to puzzles, when there is no evident feature of the world to deal with: What are the big plaques on Stegosaurus for?)

There is another stream of research in evolutionary biology which starts from an entirely different point of view, but ends up at exactly the same place. This is the study of population genetics. The idea here is to produce a description of the genetic endowment of communities on the basis of reproductive patterns and geographical distribution. The goal is to predict the rate and direction of change of genetic pools. The underlying view is still the same: the equations governing the genetic dynamics must have an optimal solution which maximizes fitness.

There has been much discussion, both within science and in popular scientific publications, about how this "classical" view of evolution has recently come under much criticism. I believe, however, that most of these discussions miss how deeply revisions have undercut the evolutionary thinking of contemporary biology.

At the very core of the matter is the question of optimality. In fact, whether at the genotypic or the phenotypic level, the classical approach is to consider separate traits which supposedly undergo progressive betterment in their fitness. But every biologist also knows that genes (or cistrones) are as intricately interrelated as are body organs, and cannot be dealt with separately. Further, the genotype and phenotype are mutually interdependent: one specifies molecular species, the other specifies which of the molecular species gets expressed. (In this sense, to speak of a genetic "program" for a species is at best misleading.) To search for paths of optimization in separate traits, given this degree of mutual specification, is to say that one tries to clamp down this interrelatedness as much as possible and hope for the best. The best is usually expressed as some sort of trade-off or compromise between traits. But even this is too feeble. The search for trait optimization has, in fact, failed to produce basic mechanisms capable of explaining major evolutionary phenomena, either at the genetic level or in morphological change. This failure has been documented in various critical discussions.[4]

The reliance on optimal adaptation is not the only way to understand organic evolution, and its alternatives are quite natural. But we need to move out from the classical framework to see that natural selection was never intended as a trait-by-trait optimization. It states, rather, minimal conditions which will be satisfied under the conditions of differential reproduction among the members of a population. This amounts to setting broad boundaries within which many pathways may be taken, as in a proscriptive rule (what is not forbidden is allowed). But this is a far cry from a prescriptive rule (what is not allowed is forbidden). Here are two concrete illustrations of what this means.

First, natural selection does not necessarily lead to steady betterment in some trait. At the genetic level, this is also true: genetic interactions do not lead to multiple combinations with other genes, all of which are phenotypically equivalent for natural selection. For example, among salamanders it is possible to find remarkable morphological constancy, which nevertheless is mediated through very different genetic pools.[5]

Second, the manifestation of genetic change in a population is to a very signifcant degree a manifestation of the internal coherence of the organisms themselves, much more so than through a selection process. In fact, genetic changes will inevitably disrupt the well-established paths of embryological development. But this is such a delicate and intricate process that single-step disruption is much less possible than radical alterations resulting in radically different phenotypes. This is, among other things, what underlies the apparent "punctuated equilibrium" which best describes the fossil record of, for example, marine invertebrates. Species mostly stay in evolutionary stasis, and when they change they do so, not in a gradualistic fashion, but by sudden jumps.[6]

These two dimensions of evolutionary change, neither of them minor, should suffice for now to illustrate that evolution is poorly described as a process whereby organisms get better and better at adapting. Rather they allow us to see that there are many paths of change, all of which are viable if there is an uninterrupted lineage of organisms. It is not a matter of the survival of the fittest; it is a matter of the survival of the fit. It is not the optimization of adaptation, but the conservation of adaptation that is central: a path of structural change of a lineage congruent with its environmental changes. This view of evolution, centered on the conservation of adaptation as a minimal condition, we call natural drift.[7]

In moving from an adaptationist view to an understanding of evolution as natural drift, we have also moved from a logic of correspondence to a logic of coherence. We have left behind the view of mirroring nature in adaptive terms, for a situation of tinkering with whatever is at hand.

A Walk through the Representationist Program and Back Again

By now, I hope, the ideas I am trying to convey are beginning to take shape in your mind, so that we may quicken the pace in this promenade through a similar conceptual landscape for the brain sciences. Briefly stated: what adaptationism is for evolutionary biology, representationism is for neuroscience.

Imagine for a moment a black-and-white television set, sitting in your living room, and try to see the color of the screen. It is gray. Now, imagine that you turn the device on, so that you see images. They will not only be gray but also black and white. The textbooks say that we see black in the absence of light, white with an intense light, and gray for the cases in between. But when the television is off, it has no way to produce a brightness on the screen through its electron beam, so we should see the screen at its blackest. In contrast, when the television set is on, however dimly, there should not be less illumination than when the device is off. Yet we all clearly agree that we see black when it is on.

In this simple example we have a capsule statement of the predominant way of thinking in neuroscience for the last fifty years. The idea is that the world has some specific features (such as light) which have a corresponding image inside, through some "mirroring" device (such as the eye), so as to produce a preception (brightness in this case). A feature of the world corresponds to a representation in the system, and this is the key for adaptive actions in the world.

The roots of this mode of thinking in neurobiology are far less clear than in the case of evolutionary biology. On the one hand, there seems to have been a tremendous influence of the newly formed engineering disciplines in the early forties. The increasingly sophisticated man-made devices were designed to handle specific forms of specifiable information, and they were successful at that. So there was nothing to prevent the brain from being a fleshy information picker. With the advent of computers, the engineering metaphor was solidly entrenched and became common sense. On the other hand, neurobiology itself began to describe sense organs as true filters detecting specific configurations in the organism's environment. In an extreme form, this became the single-cell doctrine of sensory perception,[8] which, though extreme, is not far from the sensibilities of most contemporary researchers. For this doctrine, not only perceptual items but also cognitive and motor abilities are encoded in particular kinds of neurons which stand for these performances.

The brain-as-computer metaphor, which we tend to take for granted, is, like adaptationism, nothing but one possible approach, and one plagued with problems at that. To illustrate my proposed alternative, let me return to the television set example.

It is evident in this case that black is not simply "represented" inside to correspond to a certain amount of light intensity. What then? One interesting answer is that the perception of black has to do with the relative activities in the overall retina. When we have images on the television screen, there are changes in the ratios of these relative activities, which is not the case with the uniform screen when the set is off. In other words, the perception of black cannot be studied in terms of the light falling on the retina (since we will see black at any level of illumination), but rather on the way this component of the nervous system is constructed so that some specific comparisons between light receptors are performed (out of the many conceivable ones). These comparisons establish levels of relative activity which are closely connected to the way brightness appears to us.[9]

Now, the retina is nothing but one tiny portion of this nervous system which throughout has the same characteristic of having multiple interconnections in a network, so that every state of neural activity only results in other states of neural activity, and every one of these states depends ultimately on the overall pattern of the entire brain. To make this a bit more concrete, we may contemplate the fate of the fibers reaching the brain from the retina:

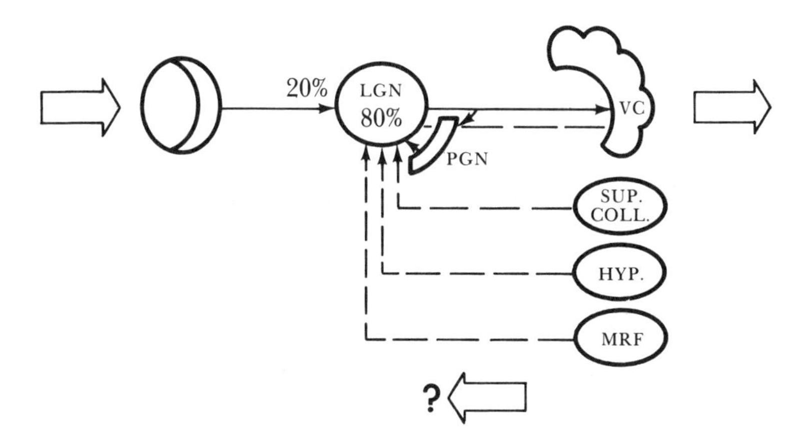

The retina projects to the brain at several places, including the thalamus at a nucleus called the lateral geniculate (LGN). The LGN is usually described as a "relay" station to the cortex. However, at closer examination most of what the neurons in the LGN receive comes not from the retina (less than 20%), but from other centers inside the brain including the visual cortex (VC), superior colliculus, hypothalamus, and the reticular formation (MRF).

What reaches the brain from the retina is only a gentle perturbation on an ongoing buzzing of internal activity, which can be modulated, in this case at the level of the thalamus, but not instructed. This is the key. To understand the neural processes from a nonrepresentationist point of view, it is enough just to notice that whatever perturbation reaches from the medium will be in-formed according to the *internal coherences* of the system. Such perturbation cannot act as "information" to be processed. In contrast, we say that the nervous system has *operational closure,* because it relies essentially on internal coherences capable of specifying a relevant world.

The differences between adaptationism and operational closure are not mere philosophical curiosities; they entail differences as research strategies. Over the last decades, the preference has been for detectors which embody particular adaptive features. The alternative is to search for cooperative mechanisms which can shape neural coherences. We cannot go further into details here.[10]

I would like to close this discussion by noting that there was nothing in the early days of modern neuroscience to indicate that it would become fascinated with representationism. As a very clear example, consider the following quotation from an important journal, as late as 1957:

> In this activity—motor sensory motor sensory . . . ad infinitum—we find a cyclical pattern, like the spinning of a wheel. . . . In the investigation of nervous activity the physiologist makes marks on the rims of the rims of the wheels of their activity. So fascinating is the process of marking . . . that the circle itself is forgotten. . . . Because of this unperceived tendency we have the scientific structure of "localization" and "representation" of functions in the nervous system. . . . [11]

Such warnings were lost, and the logic of correspondence eventually won over completely.

Autonomous Unity and Natural Drift

Let us stand back now from these two quick glances to evolutionary thought and brain science, and see them as matching pieces of a common pattern against which a new conceptual framework emerges. I can now formulate the common ground of a "new" biology in terms of the key notions presented above. This common ground can be stated in terms of two crucial changes of emphasis.

The first is putting the emphasis on the way autonomous units operate. Autonomy means here that the unit described (be it a cell, a nervous system, an organism, or a dangling mobile) is studied from the perspective of (that is, uses as a guiding thread) the way in which it stands out from a background through its internal interconnectedness. Such cooperation of self-organizing mechanisms can be made quite explicit in some cases; the research has just begun.[12]

The second change is putting the emphasis on the way autonomous units transform. Transformation means that natural drift becomes possible due to the plasticity of the unit's structure. In this drift, adaptation is an invariant. Many paths of change are potentially possible, and which one is selected is an expresion of the particular kind of structural coherence the unit has, in a continuous tinkering. Natural drift applies to phylogenetic evolution as well as to learning, depending on the unit being considered (a brain in one case; a population in the other).

I have presented a few thoughts about these ideas in the realm of the brain and evolution; clearly they can also be put to work in other realms, such as immunobiology and artificial intelligence.[13]

Autonomy and natural drift, although I have described them separately, are complementary. They are the two basic chords of the fugue I hear in the background. Let me depict them more graphically in relation to the pairs of opposites in which the classical view is rooted:

MIDDLE-WAY: META-LEVEL

	dominant view	its logical opposite
Epistemology	eternalism objectivism	nihilism subjectivism
Evolution	adaptationism	creationism
Neuroscience	representationism	solipsism
	autonomous units and natural drift: co-emergence of units and their world	

My proposal is that in this change of conceptual goggles we need to take the middle way between these logical opposites. This is not a compromise, but rather a going beyond the conflict by jumping to a metalevel.

I firmly believe that this growing framework in biology is important, as I said in the beginning, not only because it is an interesting scientific debate. It is also important because biology is the source of most metaphors in current thinking, and within biology it expresses the possibility of a world view beyond the split between us and it, where knowledge and its world are as inseparable as the inseparability between perception and action. In this middle-way view, what we do is what we know, and ours is but one of many possible worlds. It is not a mirroring of the world, but the *laying down* of a world, with no warfare between self and other. Actually, this poem by Antonio Machado says it more clearly than I could:

Caminante, son tus huellas
el camino, nada más;
caminante, no hay camino,
se hace camino al andar.
Al andar se hace camino

y al volver la vista atras
se va la senda que nunca
se ha volver a pisar.
Caminante, no hay camino,
sino estelas en la mar.[14]

(*Wanderer, the road is your footsteps, nothing else;*
wanderer, there is no pain,
you lay down a path in walking.
In walking you lay down a path

and when turning around
you see the road you'll
never step on again.
Wanderer, path there is none,
only tracks on ocean foam.)

Notes

It is a pleasure to acknowledge here my gratitude to The Lindisfarne Association and its Fellows, and to its director William Irwin Thompson in particular, for providing over many years a creative milieu where these ideas and concerns have been shaped. Financial support from the W. Woods-Prince Trust Fund is gratefully acknowledged.

1. Eskimo woman shaman poem, quoted by Rasmussen in Robert Bly, *News of the Universe* (San Francisco, Sierra Club Books, 1982), p. 257.
2. This thought experiment is inspired by the discussion in D. Hofstaedter and D. Dennett, *The Mind's Eye* (New York, Basic Books), p. 191.
3. See for example F. Jacob *Le jeu des possibles* (Paris, Fayard, 1982).
4. The most relevant recent papers for the discussion presented here are: S. Gould and R. Lewontin, *Proc. Roy. Soc.* (B) 205:581, 1979; R. Lewontin, *Reb. Sci.* 24:5, 1979; S. Gould, *Science* 216:380, 1982; and especially the crucial discussion on optimization in G. Oster and S. Rocklin, *Lecture Notes in the Life Sciences* (Providence, R.I., American Mathematical Society, 1979), Vol. II, p. 21.

5. See for instance D. Wake, G. Roth, and M. Wake, "On the Problem of Stasis in Organismal Evolution," *J. Theor. Biol.* 54:123–134.
6. See S. Stanley, *Macroevolution* (San Francisco, Freeman, 1979); J. Bonner, *Evolution and Development* (Berlin, Springer Verlag, 1980).
7. H. Maturana and F. Varela, "Evolution: Natural Drift," *The Tree of Knowledge*, (Boston, New Science Library, 1987).
8. H. Barlow, *Perception* 1:137, 1972.
9. D. Marr, *Vis. Res.* 14:1377, 1974; E. Land and J. McCann, *J. Oat. Soc. Amer.* 61:1, 1971.
10. For more details on this view of the nervous system see H. Maturana and F. Varela, *Autopoesis and Cognition* (Boston, D. Reidel, 1980); F. Varela, *Principles of Biological Autonomy* (New York, North Holland, 1979).
11. R. Goody, *The Lancet*, Sept., 1957.
12. For more on this see F. Varela, *Principles of Biological Autonomy*; F. Varela, "Self-organization" in *L'Autorganization*, Colloque de Cerisy (Paris, Edition de Seuil, 1983).
13. For the case of immunology see N. Vaz and F. Varela, *Medical Hypothesis*, 4:238, 1978. For some aspects of artificial intelligence, see F. Flores and T. Winnograd, *Understanding Cognition as Understanding* (New Jersey, Ablex Press, 1986).
14. Poem by A. Machado, from *Proverbios y Cantares*, 1930. The literal English translation is mine.

3

HUMBERTO MATURANA

Everything Is Said by an Observer

FIRST OF ALL, before I get to what I want to say about cognition, I need to point out that I am not after an explanatory principle. In part, I think that principles do not work, that whenever one has an explanatory principle, one invents a mechanism to conceal what one wants to explain. So what I propose to do is to specify a problem, and to specify also what I understand an explanation to be; then I shall discuss, from my perspective, a way of addressing the problem. In a way, I am asking you to accept as a problem what I shall propose as a problem, to accept as an explanation what I shall propose as an explanation, and, finally, to accept as an answer what I shall propose as an answer. But I am being explicit, and to make it clear that I am being explicit, I will write here in italics,

Everything is said by an observer.

And beside this I draw an eye.

Now, what is the problem? I want to think about cognition, so the problem is, then, to grasp what is the problem in cognition. I think that whenever we want to know whether or not somebody knows something about something, we ask him or her a question; and the question demands that he do something. If you want to know if somebody knows architecture, then you ask him how he would build a building, how he would proceed to make a building with certain characteristics. If he shows a way of doing this which is satisfactory to the questioner, then the questioner can say that he knows architecture. The same thing applies to biology, physics, Buddhism or religion of any kind, anything. So the problem is to identify *adequate conduct.* What constitutes adequate conduct, that is, a conduct which the questioner will accept as adequate? If I ask somebody if he knows biology, and he says, "Yes, I know biology; I am a specialist in such and such a thing," and I next ask a question to which he responds by saying or doing something which I recognize as adequate conduct in that domain, then I can say that, Yes, he knows. And I think that this is what we always do. Actually, we have no other way of assessing knowledge. Therefore, I take "adequate conduct" as an expression of knowledge. Hence, if my problem is cognition itself, or knowledge, and I recognize knowledge by seeing adequate conduct, then my problem will be to identify adequate conduct, or to show how adequate conduct arises.

What, then, would be an explanation? Usually, whenever you ask a question—ask somebody else to explain something—you expect that person to produce an answer which is satisfactory. What does satisfactory mean? It means that you don't continue questioning. When a child comes to his or her mother and asks, "Where do I come from?", the mother provides an explanation. Now, thoughout history the answers provided by mothers have changed. When I was a little boy, mothers used to tell about the bees and the flowers and things of that sort, and we children would go away to play, completely satisfied—until the next day. That was an explanation for a child, at least until the next day, when either the same question or a new one arose, because the explanation that had been given was no longer satisfactory. So the listener, then, the questioner, is the one who decides what an explanation will be: what will satisfy his curiosity? This means that if I am going to talk about cognition, I must provide an explanation which has to do with adequate knowledge, and I must be very clear about what I shall accept as an explanation.

Now, I am a biologist, a scientist, so I shall accept as an explanation only a scientific explanation. But what is a scientific explanation? Usually people think that scientific explanations have to do with predictability, that answers or propositions which allow us to predict are scientific explanations. But in my opinion this is not the case. Scientific explanations do not have to do with predictability; predictability may appear, but it is not the central point. The central point of a scientific explanation is the proposal of a mechanism. You have a question—for example, how does a horse move? A horse's movement includes trotting, and you want this explained. The scientific explanation would be a description that would imply several things, but it would have to contain a description of the mechanism which generates the horse's movements. If you want to explain lightning, you have to present a mechanism which generates lightning. This mechanism will be presented in terms of certain ideas that you have about clouds, friction, electrostatic charges, and things of this sort, but what you are actually proposing in one central idea is a mechanism that generates the phenomenon you want to explain. First, you observe the phenomenon that you want to explain, which is the question; second, you have to provide the mechanism. There is no scientific explanation if you do not propose a mechanism. But that alone is not sufficient. What is also needed in order to make the explanation a scientific explanation—and here is where the problem of predictability arises—is that the proposed mechanism generate not only the phenomenon that you want to explain, but other phenomena that you may observe as well.

Taking other observed phenomena into account is a requirement for a scientific explanation because scientists claim that what they say has something to do with the world in which we live, and that the phenomena they want to explain are the phenomena of the world. They claim that the propositions they make have a particular relation with the mechanisms that generate the phenomena because there is some isomorphism, some correspondence in structure, between the mechanisms proposed and the mechanisms in the world that generate the phenomena they want to explain. But since one can invent many possible mechanisms to generate a particular phenomenon, the scientist must select from among this multitude one in which he has more confidence because it seems to have to do with the world in which we live. This is why he also looks for some other phenomenon that will be generated by his explanatory mechanism and that pertains to the same domain as the phenomenon he wishes to explain.

So, as a scientist, I propose a mechanism. I say, "Aha! This mechanism generates this phenomenon." Of course, I have proposed the mechanism specifically because it will generate the phenomenon which constitutes my question. But then I look at this mechanism and realize that it may also generate some other phenomenon, for example, phenomenon A, which is different from the one I am explaining. That is, another phenomenon may occur in the same domain in which the phenomenon I am explaining occurs. So I look around, and if I find this other phenomenon, then I can say, "Aha! My explanation has been validated, my hypothesis has been validated. This is a scientific explanation." And that is it. It is not more, it is not less. How long will this explanation last? Until I find other phenomena which are not generated by it. At that time, I must realize that my explanation is no longer a scientific explanation. I have to drop it and invent a new mechanism which will generate not only the former phenomena but also other ones in the same domain which I consider to be important, but which are not being generated by the original mechanism. Therefore, if I want a scientific explanation of cognition, I must provide a mechanism which will generate adequate conduct—animal and human—as well as other phenomena which I can observe in the same domain. If I can do that, then, by all scientific standards, I have proposed a scientific explanation of the phenomenon of cognition—if you accept that the phenomenon of cognition is properly grasped by stating the problem in terms of adequate conduct.

What I should do next, then, is to show how adequate conduct arises in any system. This can be done, provided that we have language which is adequate to do so. First, I am going to make a couple of clarifications. An entity, anything that we can distinguish in some way, is a unity. How do we distinguish such a unity? There are many ways. For example, I could make a concrete distinction, in terms of picking it up, or a conceptual distinction, in terms of specifying a certain procedure which will carve out this unity from the background—which is at the same time specified by making the distinction. That is, when I say that something is a unity, I am also specifying all the rest of the background. This is what we do continually. If I were to ask you how many cushions there are in some room, you would count them. And in counting, you would be distinguishing cushions, performing the operation of distinction that carves these things out from the background. You may agree or disagree with someone else who is counting, but if you disagree it means that the two of you are applying different procedures of distinction. You are distinguishing different things. But if you agree, if you have the same procedure of distinction, you will count the same number of cushions, or chairs, or lamps, or whatever it is—persons, dogs, fleas, whatever. I had the honor, when I was a student at Harvard, of being the only student in a course on arthropods who knew personally fleas, mites, and all sorts of parasites. It was very interesting. I was the only one to be able to make those distinctions.

A second clarification is that we can and we do distinguish two kinds of unities—those which are simple, and those which are composite. Whenever we distinguish something as a whole and do not decompose it into parts, we distinguish it as a simple unity. Ideally the word "atom" means exactly that. If I distinguish my watch as a simple unity, then it is an atomic watch, if you wish. And the simple unity, in the moment in which you distinguish it, is specified for the operation of distinction in terms of certain properties. You can move it around, for example, or use it to point because it is long, or so on. The operation of distinction specifies or indicates the properties that characterize the simple unity. But we also distinguish composite unities. We say that the watch is made out of so many parts, things that can be separated. The components of the unity are those parts which are separate. Actually, the atom was an atom for many, many years, until the discovery of radiation allowed it to be decomposed, and then it was no longer an atom. We continue calling it an atom, but it turned out there were procedures to treat the atom as a composite unity, or a composite entity.

Now, when the unity distinguished is simple, the task is simple. One specifies properties, and that is sufficient. But when the unity distinguished is composite, there is a problem with the components, with their relations. There is a problem of composition—how are the parts put together? Here I make a distinction, which applies only to composite unities. I distinguish two features of composite unities, and I claim that we all do this. One has to do with the *organization* of a composite unity, which refers to the relations between the components that make the unity what you claim it is. For example, a chair is a composite unity. The relations between the parts that make it a chair are the organization. If I saw it into pieces and separate these pieces, would you say you still have a chair? No, you would not say that. You would say, "Why did you disorganize my chair?" I destroyed the chair by disorganizing it. The relations between components, then—that which makes a chair a chair—are its organization. A unity is a composite unity of some kind only as long as its organization is an invariant. A chair will be a chair only as long as it has the organization of a chair. If the organization changes, you no longer have a chair. This is why, by the way, I do not think I should ever use the notion of self-organization, because that cannot be the case. Operationally it is impossible. That is, if the organization changes, the thing changes. A chair is a chair, a composite unity of a particular kind, only as long as its organization is an invariant.

The second feature of composite unities has to do with *structure*. By structure I mean what most people mean by structure—the components and the relations that make a particular unity. A particular chair is made in a particular manner with particular components with particular relations between them. Another chair belongs to the same class, is a chair, is called a chair, because it has the same organization. But it has a different structure. The kinds of components that make up one chair are different from the kinds of components that make up another chair. So the organization is invariant and is common to all the members of a particular class of composite unities, but the structure is always individual. Each particular unity has a structure which realizes the organization, and which is comprised of its particular components and the concrete, particular relations that make it a particular unity. But not only that. If I were to come with a knife and secretly carve little notches on your chair, you would not ask me why I disorganized your chair; rather, you would ask why I *changed* your chair. I would have modified the chair, but it would still be a chair.

So the structure of a composite unity can be changed without its organization being destroyed. If you destroy the organization, you no longer have the unity, but something else; however you can change the structure without changing the unity in terms of its class identity, in terms of the kind of unity you have. Now, this is very interesting because we all know it. If we come home and find that our children have carved the corners off the table, we say, "What have you done to the table?", but it goes on being a table. Similarly, you continue giving the same names to your children all your life; there is something constant in the children even though they grow, and the name applies to this invariant, which is the organization, although the structure changes. Actually, in dynamic systems such as living systems, the structure is continually changing. You are changing your structure now. When I move, I change my structure, because the structure is both the components and their relations. Hopefully I can change my structure without losing my organization. As long as I can do that, or that happens to me, I am alive. But you can see that this is a very interesting situation, because when we look at things in this manner, which is what we do in everyday situations, we open an avenue to talk about change and invariance in living systems. Now, biologists know this, and when they speak about growth and evolution, they are speaking about conditions under which something remains invariant—the organization of the entity they are talking about—and something changes—the structure of those things they are talking about.

However, we still have a problem. If the explanation that a scientist proposes has to be a mechanism—and, as I have said, a scientific explanation implies a mechanism—then this explanation or hypothesis must satisfy the characteristics which make something a mechanism.

That is, it must be the description or the construction of an entity whose structure—the relations and changes of relations of its actual components—determines what happens to it. To put it another way, since a scientific explanation entails the proposition of a mechanism which will generate the phenomenon, then this proposition of a mechanism means that whatever happens to the system, which itself is being proposed by the hypothesis of a mechanism that will generate the phenomenon, is determined by its structure. It is determined by the kinds of components and the relations between the components that constitute the system. This means that whenever you have a structure-determined system, or a mechanism, and you do something to it, what happens to it does not depend on what you do to it. What happens to it depends on it. If you have a refrigerator, for example, the changes that it undergoes in its dynamic aspects do not depend on what you do to it; they depend on how it is made. We know this very well from using any of those pushbutton machines, in which if you push a button something happens—it washes, it glows, it plays music—which is not determined by your pushing the button, but, rather, is triggered by the pushing of the button.

So in structure-determined systems, in mechanisms or systems that are defined and constituted structurally, what happens to the system depends on how it is made. The interactions that the system undergoes can only trigger changes in it. You do not instruct a system, you do not specify what has to happen in the system. If you start a tape recorder, you do not instruct it. You trigger it. And living systems, if they are to be explainable, must be treated as structure-determined systems, defined by certain organizations. Hence they must be systems in which whatever happens to them is determined in them by their structure. The interactions they undergo will only trigger changes in them; they will not specify what happens to them. This is a very serious point, one that should not be taken lightly. What I am saying is that to a structure-determined system, nothing can happen which is not determined by it—by how it is made, its structure. You are forced to accept this if you want me to provide a scientific explanation of living systems, because I cannot provide a scientific explanation of systems that are not structure-determined. That is, I cannot provide a scientific explanation for systems which do not admit mechanistic experimental hypotheses. So if you want me to provide a scientific explanation which has to do with something that living systems do, such as adequate conduct, then you are asking me to treat the organism or living system as a mechanism, as a structure-determined system.

For a system to change its dynamics of state, then, for it to change what it does, even though it maintains its identity and we still call it the same name, means that its structure must change. If I have a friend who turns from Catholicism to Buddhism, his behavior will be different, so there must have been a structural change. He could not change his behavior if his structure has not changed. But his structure is changing anyway, because he is a dynamic system, so in a way that is not the problem. The problem is, what structural change took place so that he changed from Catholic to Buddhist. Our problem is indeed to explain adequate conduct, to show how adequate conduct arises. This is a problem of showing how the structure of a living system changes in a manner such that we see a particular adequate conduct which we did not see before, or we go on seeing adequate conduct even through we know the structure is changing and the medium in which the system exists is changing also. The problem is to handle the problem of structural change and to show how an organism, which exists in a medium and which operates adequately to its need, can undergo a continuous structural change such that it goes on acting adequately in its medium, even though the medium is changing. Many names could be given to this; it could be called learning. But we also have the question of how the organism originally has an adequate conduct in the place where we find it. I shall answer this question first.

Why does an organism, a living system, a person have the conduct that it has where we find it or her or him? Why am I behaving the way I am behaving? This is a question that also has to do with evolution, in that in order to understand what is taking place in evolution, one must understand what is taking place in the individual through its life history, in the ontogeny. I shall answer this question of behavior in general terms. If I have a living system—and, although I will not go into it here, I depict a living system in this manner because it is a closed system, a system which only generates states in autopoesis—then this living system is in a medium with which it interacts. Its dynamics of state result in interactions with the medium, and the dynamics of state within the medium result in interactions with the living system. What happens in interaction? Since this is a structure-determined system—and I cannot speak as a scientist if I do not treat systems in this manner—the medium triggers a change of state in the system, and the system triggers a change of state in the medium. What change of state? One of those which is permitted by the structure of the system. There are, of course, many changes of state that the structure of a particular system would permit, and the one that occurs depends on the particular circumstances. So in the interaction of a living system and its medium, although what happens to the system is determined by its structure, and what happens to the medium is determined by its structure, the coincidence of these two selects which changes of state will occur. The medium selects the structural change in the organism, and the organism, through its action, selects the structural change in the medium. Which structural change takes place in the organism? One that is determined by structure. Which structural change takes place in the medium? One that is determined by structure. But the sequence of these is determined by the sequence of interactions. The medium selects the path of structural transformation that a living organism undergoes during its life.

There are structural transformations, it is true, that result from a system's own dynamic, but those which have to do with its medium are selected through the interaction with the medium. Two organisms ideally equal in the initial state, but in different media, will undergo different sequences of interactions. Hence they will have different personal histories, individual histories, different histories of structural change. When I was a medical student, other students always went to sleep in anatomy lectures. And so the professor used to say, "Please wake your friend. I think he will be an anatomy professor when he grows up; he is sleeping now." I did not sleep in the lectures, so I never became an anatomy professor. So in the particular relation of two systems which have different structures and independence with respect to the interaction, each selects in the other the other's respective path of structural change. If this history of interaction is maintained, the outcome is an unavoidable one. The structures of the two systems will have coherent histories, although in each of them the structural changes will be determined by the structure. So after a certain history of interaction, we as observers will observe a certain correspondence in the structures of the two systems, and this correspondence is no accident. It is the necessary result of this history, the ontogeny of the individual in this medium. None of us is here by accident. All of us are here as a result of our particular histories of interactions in our media. So this congruence that one observes is no accident. This by itself, in principle, explains the most apparent features of adequate conduct. Adequate conduct is conduct which is congruent with the circumstances in which it is realized. Conduct is something that one sees, the changes of state of an organism in a medium, as seen by an observer, by the eye, this fellow who sees and describes these changes of state of the organism in its medium as conduct.

What I am saying, then, is that the life history of every organism is a history of structural change in coherence with the history of structural changes of the medium in which it exists, as realized through the continual mutual selection of the respective structural changes. The congruence between an organism and its medium is, hence, always the result of its history. This is valid for each individual, for each organism. Each organism begins its existence as a cell, and as a cell it has certain initial structures. Now the initial structure of each organism at the beginning of its individual history is itself the result of another history, which is the history of the phylogeny—the sequence of reproduction leading to that cell which is the beginning of a particular organism. And in that history of the phylogeny, the following has taken place: In each reproductive step of each life previous to an individual organism, the then existing organism reproduces at least two other organisms of the same kind, and the one that can realize itself and reach the stage of reproduction participates in the lineage. The other one—let us suppose one does not reach that stage—does not participate in a lineage. Here the participation or non-participation in a lineage, the reaching or not reaching of the next stage of reproduction, depends, of course, on whether the ontogeny is realized or not. If the ontogeny is realized, that is, if this organism lives until reproduction, it is realized only because the organism maintains invariantly its correspondence with its medium. Its structure is changing, and the medium is changing, but the coherence with the medium is maintained invariantly. Adaptation is an invariant. If adaptation were not an invariant, it would stop, and the organism would disintegrate, die.

So every cell is itself the result of a long history, which implies millions of years, a history of successive, successful reproductions, and every cell pertains to one of the many lineages which possibly come from one common point in some faraway past. But along this history, the phenomenon of the organization of the cell, the condition of living, has remained invariant, and adaptation has remained invariant. Structures of the organism have been changing as a result of a continuous selection through the structural changes, through the interactions of the organism with the medium. So not only are we here now as a result of our personal histories, but we are here now as a result of the history of our ancestors. In a way, we all have the same age, and all our cells have the same age—millions of years—if we see not only our own individual ontogenies, but also the phylogenies, the history which is responsible for the structural changes that have led to our particular kind of coherence. That particular kind of coherence appears expressed in adequate conduct.

Now, I realize that you may think that there is a trick in this problem of adequate conduct, so perhaps I can illustrate the idea by relating an interesting anecdote which I read in *Time* magazine some years ago. A young student had to take an examination in physics. The professor handed him an altimeter and told him to determine the height of the campus tower with the altimeter. The student went to Woolworth's, bought a piece of string, went up the tower, tied the altimeter to the string, lowered it to the base of the tower, and then measured the string: 32 meters, 50 centimeters. Flunk. But the student appealed, and the commission of education, or whatever, decided that he had the right to take the examination again. So the professor handed him the altimeter and told him to determine the height of the tower with it. This time, the student got a goniometer, which measures angles, went to some distance from the tower, and used the height of the altimeter to triangulate the tower. Flunk. Again petition, again concession, again determine height of tower with altimeter. Now, the tower happened to have a beautiful helicoidal staircase, so the student went along each step with the altimeter, determined the path of the screw, and again came up with a figure. This perverse student invented seven ways of determining the tower height without reading the altimeter! Of course, the question is, did he know physics or not? Did he have an adequate conduct? When the professor flunked him, it appeared that he did not have an adequate conduct. He failed to show adequate conduct under the circumstances in which the question was asked. So if the crucial thing was the opinion of the teacher, he failed. But the education commission had a different opinion, and so he did not fail.

Now, the teacher that determines this fundamental adequate conduct for us is life. If we remain alive, we have adequate conduct—whichever way we manage to remain alive. And if we reproduce, we participate in a lineage. However, if the criterion is determined by the professor, we have adequate conduct only to the extent that we satisfy the demands of the professor. Can what I have said in very general terms that obviously apply to our fundamental call, which is to live, also be used to explain adequate or inadequate conduct in front of the professor? Yes, and I shall show you how. Suppose that instead of considering only the medium which I have previously posited—an inert physical medium, something that we would not call living—I put in another living system. Then the situation will be this: We shall still have the previous interactions taking place, but other interactions will also take place. But my argument about interactions will still apply for these new interactions, because the phenomenon of selective interaction, of selecting the structural change in the other, does not depend in any way on the characteristics of the agent with which the change is made, provided interaction takes place. In fact, the organism specifies what it admits as an interaction. Each one of you specifies what you admit as an interaction. To other things you are transparent. You don't understand what I say when I use an unknown language; you specify what languages you understand.

So there is no restriction on what other things one can interact with, but if it happens that the other entity is a living system, then we have an adaptation that involves another living system. And the invariance of adaptation involves another living system. When this occurs, I claim, although I shall not go into its full development, that we have a *linguistic* domain. Whenever we have organisms that through a history of interaction continue interacting with each other, we have a linguistic domain. But notice that adaptation, invariance of adaptation, is a structural coherence, meaning that the structure of the system can be described as having mutual correspondence in a dynamic manner. I have called this *structural coupling*. The same thing happens between organisms. If there is coherence in the history of interaction, they are mutually adapted. And they will continue interacting with each other as long as there is coherence, as long as they remain mutually adapted, because each interaction will result in the selection of a particular structural change. And again, whenever this takes place, a linguistic domain is established. If this linguistic domain allows for a recursion in linguistic interaction, then we have a language, but I shall not go into that. Certainly when a professor and a student have a history of interaction, the adequate conduct of the student will reveal a coherence in the domain of interaction with the professor. If such coherence is interrupted at some moment, then the student will not have adequate conduct in the eyes of the professor. But the professor and the student do select in each other the path of structural changes as long as they maintain the relation.

To the extent that I have shown you the mechanism by which adequate conduct is generated, I have answered the question I proposed about the phenomenon of cognition. Remember that I did not ask, what is cognition? I only asked, under what circumstances do we recognize cognition? I have shown the circumstances that generate phenomena in which we recognize cognition, but I have also done something else. I have made an identity between cognition and living, at least in the absolute general terms that have to do with us as living systems. There are other, much more restricted domains of cognition, and, with respect to those, I have said that in any given domain of concentrality that we establish with another organism, the other organism will observe in us cognitive behavior—will observe in us adequate conduct. The phenomenon of cognition, from what I have said, is necessarily relative to the domain in which one observes structural coherences which are the result of the histories of interactions of the organisms.

Finally, I will present a couple of interesting ideas, although I will not be able to develop them here. When you have language, what you have is the possibility of behavior that the observer can describe as recursions in a linguistic domain of consensus. These recursions can take place because there is a very interesting peculiarity in the nervous system. The nervous system is a closed system, a closed network of components which interact with each other, and in which the dynamics of state is a continuous change of relations of activity that generates relations of activity in the same network. Which relations of activity and which changes of relations of activity take place? Those that are determined by the structure of the nervous system. Therefore, one can show that in terms of description, because the description is conduct in this linguistic domain of mutual coherences, language is not in the brain or in the nervous system, but rather in the domain of mutual coherences between organisms. When the observer observes that this takes place, and that the distinctions realized here can be recursive, can be distinctions on distinctions in this domain, then we have a language. But that can only occur because everything is taking place in a closed system. For the system, for us in our nervous system, the act of picking up a piece of paper is a particular series of changes of relations of activity in our nervous system. To drink water is again another series of changes of relations of activity in our nervous system. To speak is yet another. From the point of view of what takes place inside the organism, everything takes place inside the organism in a closed manner. But for the observer, the coherences appear as language, or linguistic interaction, things of that sort. And this is what allows, finally, for your own dynamics of state in a linguistic domain to operate as a selector of your dynamics of state.

4

JAMES LOVELOCK

Gaia

A Model for Planetary and Cellular Dynamics

MOST OF US were taught that the composition of our planet could adequately be described by the laws of physics and chemistry. It was a good, solid Victorian view, and even if you have forgotten the details you will surely have been left with the idea that anything you need to know about the Earth can be found in the appropriate textbook, if only you can find time to read it.

In a similar way, the climate was said to be a natural consequence of the Earth's position in space around that great and constant radiator, the Sun. To explain the climate anywhere on Earth was simply a matter of balancing the heat received from the Sun by the different climatic zones against the loss of heat by radiation into the cold depths of space.

On this reliable and predictable planet of the hard-core geologists, the biosphere was regarded as a by-stander or a spectator and was not permitted to enter the game. We and the rest of life were told that we were amazingly fortunate to be on a planet where everything is and always has been so comfortable and well suited for life.

I am speaking here partly in my capacity as a sort of shop steward for the nonhuman segment of the biosphere. On behalf of my members, I want to put it to you that this exclusion of life from its proper place in the running of this planet was a diabolical liberty. We think that conditions on the Earth are just right for life because we and the rest of life, by our struggles, have made and kept it so.

This is not new. The idea that life might have the capacity to mold the conditions of the Earth and optimize them for the situation of the contemporary biosphere has been hinted at in the past, notably by Redfield, by Hutchinson, and by Lars Gunar Sillen. In their times, though, it was considered so radical a thought as to be beyond discussion in mainstream science.

The earliest reference that I have found to the idea that life might have molded the Earth to suit its own needs is in the June, 1875 issue of *Scientific American,* in an article largely on the controversy concerning evolution:

> A popular, illogical dogma declares that life is the grand object of Creation; that the composition as well as the contour of the Earth's surface has special reference to its habitability; and that all things show a ruling design to fit the world to be the home of sentient creatures, more especially, of man. Strictly speaking, science has nothing to do with these dogmas. It has no means of discovering the ultimate purpose of things and no time to waste on their discussion. Nevertheless, it is difficult sometimes not to take an indirect interest in the claims of those who presume to decide such questions, at least so far as to notice how actually the facts of nature contradict their assertions. Thus, in the

> present case, it would be much easier to sustain the contrary thesis; namely, that so far from being made what it is that it might be inhabited, the Earth became what it is through being inhabited. In short, life has been the means to, not the end of, the Earth's development.

It was not until quite recently that new subjects like biogeochemistry appeared on the scientific scene. This new approach to the understanding of our Earth did not come from enlightenment in the earth sciences; instead it was inspired by the investigation of the other planets, especially Mars.

My part in this story began in 1965 when, with a colleague, Dian Hitchcock, I was working at the Jet Propulsion Laboratory in Pasadena, California. We had been given the task to examine critically the life-detection experiments then proposed for Mars.

At that time, and it now seems long ago, it was generally believed that there was a sporting chance of finding life on that planet. In any event, it was felt that the discovery of life anywhere outside the Earth would be a momentous event that would so enlarge our view of the Universe and of ourselves as to be well worth the cost of trying. Hitchcock and I did not disagree with these noble sentiments, but we were concerned that most of the experiments then proposed were much too geocentric to succeed even if there were life on Mars.

It seemed as if the experiments had all been designed to seek the sort of life each investigator was familiar with in his own laboratory. They were seeking Earth-type life on a planet not in the least like the Earth. To Dian and me, it seemed that we were guests of an expedition to seek camels on the Greenland icecap or of one to gather the fish that swam among the sand dunes of the Sahara.

I wondered if it would be possible to design a more general form of life-detection experiment, one which would recognize life, whatever its form might be. One possibility would be to look for inconsistencies in the chemical composition of the planetary atmosphere and surface to see if substances or processes were present which were inexplicable on the basis of inorganic chemistry. The idea behind this was that if the planet did bear life, that life would be obliged to use the atmosphere as a source and depository for raw materials and also as a convenient medium for the transport of its products. Such a use of a planetary atmosphere would be revealed by changes in its chemical composition, which were very improbable as a consequence of the random processes of nonliving chemistry. It would be a way of looking at Mars which made very few assumptions about the details of the life, if it were there.

As long ago as 1965, and long before any spacecraft had moved near Mars, there was nevertheless a great deal of information available about its atmospheric composition. This came from astronomical observations using a telescope tuned to infrared rather than visible radiation. The telescope was equipped with a device called a multiplex interferometer, invented by my colleague Peter Fellgett, which had the capacity to provide an exquisitely detailed analysis of the gases in the planet's atmosphere. This powerful system was used by Pierre and Janine Connes at the Pic de Midi Observatory in France, and it revealed that the Martian atmosphere was one dominated by carbon dioxide and apparently very close to the state of chemical equilibrium. According to our theory, Mars was a planet very unlikely to bear life.

To check this prediction, we needed a planet which did have life, and, of course, the only one available to us was the Earth. It was not difficult for us to set up a *Gedanken* experiment with an imaginary infrared telescope on Mars. Such an instrument looking back at Earth could easily have found the presence and the abundance of the gases oxygen, water vapor, carbon dioxide, methane, and nitrous oxide. From this information, together with that of the intensity of sunlight at the

Earth's orbit, it is possible to deduce with near certainty the presence of life on Earth.

The argument goes as follows: We have an abundance of oxygen, 21% of the atmosphere, and a trace of methane, at 1.5 parts per million. We know from chemistry that methane and oxygen will react when illuminated by sunlight, and we also know the rate of this reaction. From this we can confidently conclude that the coexistence of the two reactive gases methane and oxygen at a steady level requires a flux of methane of 1,000 megatons a year. This is the amount needed to replace the losses by oxidation. Furthermore, there must also be a flux of oxygen of 4,000 megatons a year, for this much is used up in oxidizing the methane. There are no reactions known to chemistry which could make these vast quantities of methane and oxygen starting from the available raw materials, water and carbon dioxide, and using solar energy. Therefore, there must be some process at the Earth's surface which can assemble the sequence of unstable and reactive intermediates in a programmed manner to achieve this end. Most probably this process is life.

We had proved our method and used it to show that Mars was probably lifeless. Needless to say, this was not welcome news to our sponsor, the National Aeronautics and Space Administration. They badly needed reasons to go to Mars, and what better than to find life there? Much worse, it was hardly good publicity for NASA to claim that work they had funded proved that there was life on Earth. It would have been a gift to Senator Proxmire, and I was not surprised to find myself soon unemployed.

When I returned to England in 1966, the thought kept recurring: How is it that the Earth keeps so constant an atmospheric composition when it is made up of highly reactive gases? Still more puzzling was the question of how such an unstable atmosphere could be perfectly suited in composition for life. It was then that I began to wonder if it could be that the air is not just an environment for life but is also a part of life itself. To put it another way, it seemed that the interaction between life and the environment, of which the air is a part, is so intense that the air could be thought of as being like the fur of a cat or the paper of a hornet's nest: not living, but made by living things to sustain a chosen environment.

An entity comprising a whole planet and having the powerful capacity to regulate its climate and chemical composition needs a name to match. I was fortunate in having as a near neighbor at that time the novelist William Golding. When I discussed this with him during a walk around our village, he proposed the name *Gaia*, that which the Greeks used for the Earth. I was glad and grateful, for it was a simple four-letter word and not an acronym for one of these ugly phrases so beloved of my fellow scientists. Somewhat naively, I imagined that it might preempt the use of such as chemico-bio-geo-cybernetics or worse. It did not, of course. Nevertheless, when I use the word Gaia from now on, it is the name for the hypothetical system which regulates this planet.

In the later 1960s the only scientists to take Gaia seriously were the eminent Swedish geochemist Lars Gunar Sillen and the equally eminent American biologist Lynn Margulis. Lynn and I have collaborated ever since in its development, and the evidence we have gathered falls into two categories.

First there is the thermodynamic evidence, the kind of evidence I have already mentioned in connection with the coexistence of oxygen and methane. It is concerned with the extent to which the present real Earth is recognizably different from an Earth made of the same material and in the same position in the solar system, but which does not bear life. This difference can be measured in terms of the extent to which the chemical composition of the soil, oceans, and air differ from the predicted equilibrium state. The difference is a measure of the reduction in entropy due to the presence of life.

To illustrate this difference, let us consider something of the composition of the atmospheres of a number of planets: Venus, Mars, Earth, and Jupiter, and also our hypothetical Earth, which somehow has had all of the life wiped out, but is otherwise exactly the same as the real Earth—whatever drastic event caused life to be wiped out was specifically biocidal and did not do anything to the chemistry or to the climate at the moment it happened. Now, what matters in an atmosphere is not how much of a gas is there, but rather how much of it is flowing through the atmosphere. In our atmosphere we have 80% nitrogen, but it is a fairly inert gas and doesn't flow through the atmosphere as rapidly as does methane, which is present only in the amount of 1.5 parts per million. There are three important classes of gases present in planetary atmospheres: oxidizing gases, such as oxygen and carbon dioxide; neutral gases, such as nitrogen and carbon monoxide; and what the chemists call reducing gases, such as methane, hydrogen, and ammonia. In general, the oxidizing and reducing ones want to react with one another, and usually quite vigorously.

Now, the point of this illustration is that the atmosphere of the two lifeless "terrestrial" planets, Venus and Mars, contain just oxidizing and neutral gases, whereas those of the great gas giants, of which Jupiter is a good representative, have nothing in them but the reducing gases. The Earth, our living Earth, is quite anomalous; its atmosphere has the reducing gases and the oxidizing gases all coexisting—and this is a most unstable situation. It is almost as if we were breathing the sort of air which is the premixed gas that goes into a furnace or into an internal combustion engine. Ours is a really strange planet. Now the hypothetical sterile Earth would have an atmosphere just like that of Mars and Venus: oxygen would be a mere trace of what it is now on Earth; nitrogen would be gone largely, into the seas; and methane, hydrogen, and ammonia would vanish in just a few years.

When the air, the ocean, and the crust of our planet are examined in this way, the Earth is seen to be a strange and beautiful anomaly. The evidence that Lynn Margulis and I, and others, especially Michael Whitfield, have gathered over the years establishes almost beyond doubt that the Earth is a biological construction. All of the compartments of the Earth's surface are kept at a steady state, far removed from the expectations of chemistry, through the expenditure of energy by the biosphere. The next step is to establish that this construction is optimized for the contemporary biosphere. There are reasons to expect that buried in the thermodynamic evidence is the information needed to establish Gaia's existence as a control system. There is as yet no formal physical description of life itself, and it may be that the same formalism is needed to prove Gaia.

There is another way too of approaching Gaia and that is through cybernetics. The usual way to examine an hypothesis cybernetically is to compare the behavior of the real Earth with that of a dynamic model. Robert Garrels and his colleagues have done this for the cycles of some of the major elements which flow through the surface compartments (the oceans, the crust, and the atmosphere) of the Earth. When they considered the effects of the presence of life upon this flow, they concluded, to quote Garrels, "The Earth's surface environment can be regarded as a dynamic system protected against perturbations by effective feedback mechanisms." In a similar way, Michael Whitfield has examined the cycles of elements in the oceans and concluded that machinations by living things plays a major part in the distribution and abundance of the various elements dispersed in the sea.

Another way of examining the Earth cybernetically is to ask the question: What is the function of each gas in the air or of each component of the sea? Outside the context of Gaia, such a question would be taken as circular and illogical, but from within it is no more illogical than asking: What is the function of the hemoglobin or of the insulin in the blood? We have postulated a cybernetic system; therefore it is reasonable to question the function of the component parts.

Let us consider the gases of the air in this way. Oxygen is the dominant gas even if not the most abundant. It sets the chemical potential of the planet. It makes it possible, given something combustible, to light a fire or power a combustion engine anywhere on the Earth. It makes it possible for birds to fly and for us to think.

Any functional component of an active system is likely to be regulated; with an important and powerful component like oxygen, the need for regulation must be great. What evidence have we that oxygen is regulated? Certainly for several hundreds of millions of years it can not have been more than a few percent less than now, or the larger animals and flying insects could not have lived. My colleague Andrew Watson has demonstrated in some elegant experiments that it can never have been more than 4% greater than now, and probably not even 1% more. His experiments showed that the probability of forest fires is critically dependent upon oxygen concentration and that a mere 1% rise in oxygen increases the probability of fire by 60%. At 25% oxygen even the damp detritus on the floor of a rain forest could be ignited by lightning. Once lit, the forests would burn in an awesome conflagration more fierce than we have ever known. If this atmospheric oxygen content of 25% were long sustained, all standing vegetation would be burnt from the Earth's land surfaces. This clearly is a situation that is far from optimum. Our present oxygen level of 21% is a nice balance between risk and benefit; fires do take place, but not so often as to offset the advantages that a high potential energy gives.

Although this is not the place to describe our work in this area, we have continued to investigate oxygen regulation and think now that an otherwise enigmatic, apparently wasteful process of the biosphere—that of producing methane, only to have it flow up into the atmosphere where it is oxidized, apparently doing no good—is in fact part of a feedback loop concerned with the regulation of oxygen. If this is true, methane has an important function. Similar arguments can be used to assign functions to the other gases of the atmosphere, even nitrogen.

One of the most convincing arguments for Gaia comes from the apparent need for regulation of the climate. Although it is common knowledge among astronomers, it is not generally well known that our sun is warming up exponentially, and has done so since the origin of the planet. The rate of rise of the output of the sun is such that the output is likely to have increased by between 30 and 50% since life began. It is a property of stars to increase their output of heat and light as they grow older, and there is no reason to suppose that our sun is an exception. Obviously the climate at the start of life would have to have been equable, neither freezing nor boiling. Now, the course of the temperature during the time that life has existed is not known for certain, but all evidence indicates that it has stayed remarkably constant. A 30% increase of solar output above the present level would bring us to the boiling point, so if the present rate of increase of solar output has occurred since the beginning of life, why are we not boiling now?

Sagan and Mullen were the first to offer a plausible resolution. They suggested that the young Earth had an atmosphere rich in ammonia and that this gas, through its capacity to absorb outgoing infrared radiation, acted as a blanket which kept the planet warm in spite of the cooler sun. Others not liking ammonia have proposed that 5% to 10% of carbon dioxide could achieve the same result.

Intimations of Gaia come from the realization that for the Earth to have developed from its origin to the present day required a smooth and continuous decrease of whatever the blanket gas was which kept it warm, so that the thickness of the blanket matched the increasing warmth of the sun. Ingenious and even plausible schemes have been proposed in which, for example, the rate of weathering of the rocks always removes carbon dioxide at just the right rate so that the planet stays at a stable temperature. These schemes lose credibility when we consider the fact that the Earth's climate is poised between two more stable but quite lethal climatic regimes, one freezing, the other near boiling. In addition, when the natural tendency of infant life to eat the blanket is also taken into account, the life's survival unscathed for all those long years does seem to be a persuasive indication of Gaian regulation.

Nothing more than chance may be needed to explain any one of the items of evidence I have mentioned, but when all are taken together as an ensemble, and especially when the known constancy of the Earth's environment is taken into account along with the certain knowledge that many major perturbations have been endured, then a closer look at Gaia does seem worthwhile.

In its development as an hypothesis, Gaia has been neglected rather than criticized by the scientific community. Geochemists have preferred to believe that while some changes in the Earth's composition may be attributable to the biosphere, such changes are passive and do not in any way constitute regulation. The only direct criticism so far has been that of molecular biologists, most clearly expressed by Ford Doolittle, who maintains that there is no way by which Darwinian natural selection could lead to a quasi-immortal entity like Gaia. Selfish genes could never form so altruistic an association. We take this criticism seriously but disagree, at least on the grounds that it is based upon the false assumption that adaptive evolution occurs independently of the environment in which adaptation occurs. In fact, every evolutionary step of a component of the biosphere has the capacity to change the environment. Sometimes, as with the first appearance of atmospheric oxygen, the change is a drastic one indeed. When the environment is changed by a new speciation, then adaptation is forced upon many others, and so change continues.

Such a process is familiar to mathematicians who use numerical methods. It is that of iteration, wherein a sequence of guesses converges upon the unattainable truth. More often than not, such processes lead to the minimization of change and a new stability.

Those, then, are some of the items of evidence and criticisms of Gaia. If I had presented all of them, it would still only corroborate and not prove her existence. In any case, in science it is usually less useful to sanctify an hypothesis than to use it as a sort of looking glass through which to see the world differently. So let us for a moment suppose that Gaia does exist, and see what are the consequences of her presence on our current concerns.

The first thing to come to mind is the effect of the expected growth in numbers of the human species, together with that of its dependents in the way of crops and livestock. Together we consume an increasing proportion of the total material resources of the Earth. What are the consequences of this with or without Gaia?

Those environmentalists who believe that the composition and climate of the Earth are independent of the biosphere regard life as fragile and in danger of destruction. I do not disagree; if life and the environment were evolving independently, then life would be fragile, for it would be at the mercy of any adverse change.

There is a strangely familiar ring to the word "fragile." It was widely used in Victorian times to describe women, possibly to justify male domination. Whenever an environmentalist tells me that life on Earth is fragile and may fall apart if, say, the ozone layer is slightly depleted, I think of my Victorian grandmother. If we accept Gaia, at least for argument, this fragility is nonsense. Gaia, like the Victorian women, is very tough indeed. Like them, she has had to be to endure the insults.

Just how tough life or Gaia is, is proven by its survival in spite of at least 30 near mortal blows from the impact of planetesimals. Every 100 million years or so, a small planet about twice the size of Mount Everest and moving at 60 times the speed of sound hits us. The kinetic energy of its motion is so great that if it were uniformly dispersed over the whole Earth it would be equivalent to the detonation of 30 Hiroshima-sized atom bombs for every square mile. Fortunately, its effects are to some extent localized.

An impact such as this 65 million years ago caused the extinction of over 60% of all species then present. It was one of at least 30 such impacts since the start of life, and some were 20 times more severe. Gaia can hardly be fragile to withstand blows like these, and, indeed, the surge of speciation following such events is indicative of her capacity to recover. It is even possible that we as a species resulted from the stimulation of one of these recent impacts.

It seems very unlikely that anything we do will threaten Gaia. But if we succeed in altering the environment significantly, as may happen with the atmospheric concentration of carbon dioxide, then a new adaptation may take place. It may not be to our advantage.

When we talk about life or the biosphere, we tend to forget that procaryotes, simple bacteria, ran a successful biosphere and represented life on Earth for nearly 2 aeons (two thousand million years). They are still today responsible for a great deal of the running of the present system. Lynn Margulis once remarked that the true function of mammals, including humans, might be to serve as ideal habitats for the few pounds of bacteria carried in the guts. They are kept warm and well fed there, in what must seem their own private heaven. Such thoughts of Gaia remind us also that there is more to life than humans, cuddly animals, trees, and wild flowers. Those who rightly are concerned for these must also be concerned for their less attractive infrastructures.

A frequent criticism of the Gaia hypothesis is that it engenders complacency through the belief that Gaian feedback will always protect the environment against any harm that mankind might do. It is sometimes more crudely put that Gaian ideas give industry the green light to pollute at will.

Scientific hypotheses are all too often used as metaphors in arguments about the human condition. This misuse of Gaia is as inappropriate as was the use of Darwin's theory to justify the morality of laissez-faire capitalism. Gaia is an hypothesis within science and is therefore of itself ethically neutral. We have tried hard to keep faith with the rules of science. If the hypothesis is used outside this context, I will say again that it is just a looking glass for seeing things differently. It is all too easy with a looking glass to reflect yourself accidentally.

The environmentalist who likes to believe that life is fragile and delicate and in danger from brutal mankind does not like what he sees when he looks at the world through Gaia. The damsel in distress he expected to rescue appears as a buxom and robust man-eating mother. The same environmentalist will take the Second Law of Thermodynamics as a mirror and see in it a justification for the apocryphal Murphy's Law, "If anything can go wrong it will." He views our universe as the setting for a tragedy, with us as the players in a deadly game in which we cannot break even, let alone win.

I see through Gaia a very different reflection. We are bound to be eaten, for it is Gaia's custom to eat her children. Decay and death are certain, but they seem a small price to pay for life and for the possession of identity as an individual. It is all too easily forgotten that the price of identity is mortality. The family lives longer than one of us, the tribe longer than the family, the species longer than the tribe; and life itself can live as long as it can keep this planet fit for it.

Perhaps the strangest knowledge to come from our quest for Gaia has been the realization that, robust though she may be, the conditions of our Earth are moving close to the point where life itself may be not far from its end. The quite unstoppable increase of the sun's heat soon will be beyond the capacity of regulation or adaptation. In human terms the Earth is still forever inhabitable. But in Gaian terms, if the length of life on Earth were one year, we are now in the last week of December.

Before our Earth becomes a problem in planetary geriatrics, with flimsy contraptions in space like sunshades to keep it alive a few more millennia, I hope that the parallel moral problems inherent in human geriatrics are solved.

It is only pessimistic to see our Earth, as well as the universe itself, running down to a heat death if you are one of those who wants to have their cake as well as to eat it. You can't use a flashlight to see your way in the dark and expect also to have the batteries last forever. It is the running down of the universe that made the Earth possible, and the sun, and it is the running down of the sun that has made life and us possible. It has to end sometime.

5

LYNN MARGULIS

Early Life

The Microbes Have Priority

FOR MOST OF the history of life on this planet, the living landscape resembled Gustave Courbet's time-forgotten seashore expressed in his painting *Marinescape*. Although inconspicuous, life in the form of bacteria and their diverse communites changed forever the surface and atmosphere of the planet. Although tiny, early life was complex and original. In mud flats, evaporite expanses, fens, and ponds, microbes evolved innovations that we now associate with animals and plants: reproduction, predation, movement, self-defense, sexuality, and many others.

Are the well-formed filaments found so recently in the Warrawoona Series of northwestern Australia really evidence of the oldest life on the planet? Do the fossils found in the great Gunflint Iron Formation of Ontario tell us that bacteria were instrumental in the accumulation of the most important iron reserves in the world? These questions have not yet been resolved, but even in raising them now we are beginning to realize that life is a much more dynamic part of "geophysiology" than we had suspected before.

The Earth has had a solid surface of rocks for about four billion years. The oldest fossils—of microscopic isolated spheres resembling modern bacteria—are about 3.5. billion years old. Yet until about half a billion years ago, no large multicelled organism—no animals or plants—inhabited the Earth. At about that time, the fossil record shows, marine animals appeared all along the world's seashores. From these animals and the seaweeds that fed them have descended many forms of life. Since then, life has crawled out on the land, flowering plants have appeared and become the dominant vegetation, and all the insects, fishes, reptiles, birds, and mammals have appeared. The history of human beings is a mere moment compared with what went before—the first modern human remains, those of *Homo sapiens sapiens*, appear in the fossil record of only about 35,000 years ago.

Is evolution going faster and faster? Why did it take three billion years for the elaboration of the single cell into the large multicelled organism? These single-celled organisms invented the chemical and biological strategies that made more intricate life forms possible. During those first three billion years, the cell went through profound evolutionary development; it was engaged, quite literally, in evolving its working parts. By the time marine algae and animals appeared, microbes had developed all the major biological adaptations: diverse energy-transforming and feeding strategies, movement, sensing, sex, and even cooperation and competition. They had invented nearly everything in the modern repertoire of life except, perhaps, language and music.

Until recently, most efforts at reconstructing the ways in which organisms have evolved were directed toward animals and plants. That the simpler but more abundant and diverse microorganisms are also products of a long evolutionary history is a new realization, one that has developed from recent discoveries in several fields, including microbiology, biochemistry, and geology. Perhaps the most illuminating discoveries have been made by use of the electron microscope, which, using an electron beam instead of light, can magnify as much as 500,000 times. Organisms thought to be similar have turned out to be full of surprising differences; structures and organisms apparently quite diverse have turned out to have a great deal in common. Knowledge of the detailed fine structure of cells has led directly to insight into evolutionary relationships.

Much evidence for evolutionary history is gained by studying extant organisms. Fortunately for students of ancient life, successful innovation perpetuates itself; once complicated patterns of growth and metabolism arise and thrive, they tend to persist. The minuscule bacteria have become optimally adapted to such ancient and persistent niches as rocky seashores, mud flats, stream beds, and salt flats. By studying these patterns of metabolism, gas exchange, and behavior in these ubiquitous cells, researchers have begun to piece together a picture of what their earliest ancestors were like.

The cells of large organisms, such as plants and animals, generally are larger than bacterial cells. They also differ in other fundamental ways. Animals and plant cells always contain organelles, distinct intracellular organs that differ recognizably from their surroundings in the cell. One organelle that they all have is the nucleus. Separated from the rest of the cell by a membrane, the nucleus is a bag that contains the genetic material, deoxyribonucleic acid (DNA), as well as crucial large protein molecules and ribonucleic acid (RNA). By definition, a cell that contains its DNA in a membrane-bounded nucleus is a eukaryote. The living world is unambiguously divisible into eukaryotes and prokaryotes, cells that lack nuclei. All large and elaborate forms of life are composed of eukaryotic ("truly nucleated") cells, whereas bacteria and their microbial relatives are composed of prokaryotic ("pre-nucleated") cells.

In eukaryotic cells, the DNA is tightly coiled with protein into chromosomes, rodlike bodies inside the nucleus. The DNA of prokaryotes, in contrast, is a single

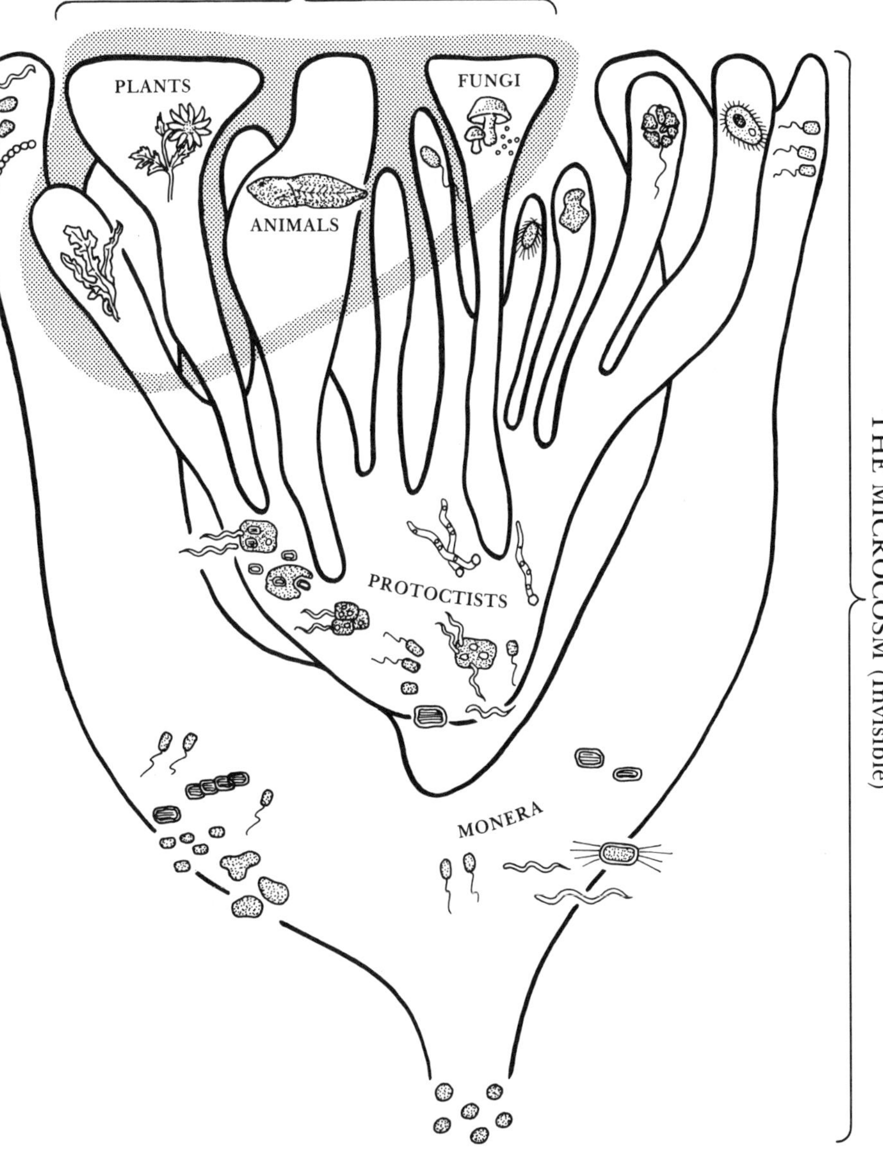
THE MACROCOSM (Visible Organisms)
PLANTS
FUNGI
ANIMALS
PROTOCTISTS
MONERA
THE MICROCOSM (Invisible)

long circular molecule of DNA that floats free in the interior of the cell. With very few exceptions, all eukaryotic cells contain mitochondria, membrane-bounded organelles in which oxygen is used to "burn" food molecules that provide energy for most other cell activities. Another organelle that generates energy is the chloroplast, a membrane-enclosed unit that contains chlorophyll. Cells of green plants and green algae contain at least one and as many as hundreds or even millions of chloroplasts. They are the sites of photosynthesis, the process by which cells transform the energy of sunlight into chemical energy. In prokaryotes, the consumption of food molecules and the process of photosynthesis are not confined to enclosed organelles, but take place on membranes distributed widely through the cell.

Motile eukaryotic cells typically carry on their external membranes short hairlike structures (cilia) or longer whiplike structures (flagella). Both cilia and flagella are made of bundles of small hollow microtubules arranged in the same elaborate pattern; for this reason they are both called undulipodia. The beating of undulipodia can move the cell itself or can move particles and fluids past a stationary cell. Among prokaryotes, the analogous structures (called flagella) are far smaller and simpler—they are single stranded. Other components unique to eukaryotic cells are centrioles, small dotlike bodies that appear during cell division; vacuoles, membrane-enclosed spaces that take part in fluid and salt regulation; lysosomes, small packages of chemicals that break down food particles for intracellular digestion; and Golgi bodies, groups of flattened membranous sacs that package and transport products synthesized by special cells. Golgi bodies are especially conspicuous in cells that produce hard shells, skeletons, or glandular secretions such as digestive juices.

The earliest life on Earth consisted only of bacteria-prokaryotic cells. Organisms made of eukaryotic cells did not appear on the scene until much later. Precisely when this evolutionary innovation took place has been the subject of much debate. Eukaryotic cells may be more than two billion years old, but they cannot be less than about 700 million years old; by that time, marine animals made of such cells were distributed along many seashores. How did eukaryotic cells arise? The sequence of events linking prokaryotic ancestors with their eukaryotic descendants is the subject of wide discussion, and different hypotheses—the subjects of many laboratory investigations—have been put forward. The theory I favor is that certain organelles of the eukaryotic cells originated by symbiosis.

Symbiosis can be defined as the intimate living together of two or more organisms, called symbionts, of different species. According to the symbiotic theory of the origin of the eukaryotes, once-independent microbes came together, first casually as separate guest and host cells, then by necessity. Eventually, the guest cells became the organelles of a new kind of cell. Such a sequence of events can be found in the symbiotic relationships between many modern life forms. Many organisms live inside, on top of, or attached to other organisms. Hereditary symbioses—those in which the partners remain together throughout their life cycle—are surprisingly common. In some instances, one partner can manufacture its own food by photosynthesis, but the other cannot. Organisms of the first type, able to capture the Sun's energy directly and use it to synthesize the compounds they need for growth and reproduction, are known as autotrophs ("fed by self" from Greek *trophe*, nourishment). Organisms of the second type are called heterotrophs ("fed by others").

Lichens are a common example of a symbiotic relationship. Characteristically flat, crusty, plantlike organisms that can survive in alternately wet and dry and harshly cold environments, lichens are symbiotic partnerships between algae (autotrophs) and fungi (heterotrophs). The algal cells are enfolded by tough threadlike fungal cells, which protect the algae from the harshness of their environment. The algae, which must live in water when they live independently, produce food photosynthetically for themselves and for their fungal partners.

Certain bacteria that inhabit the mud of lake bottoms also enter into symbioses. A larger partner that can swim will team up with several smaller, immobile forms that can produce their own food by photosynthesis. This consortium of bacteria then swims as a unit which has the capabilities of both partners. Coral reefs, too, depend on the association between tiny coelenterates (corals, jellyfish, sea anemones) and their symbiotic partners, typically single-celled dino-mastigotes of the genus *Symbiodinium*. The dino-mastigotes, which live inside the cells of their hosts, photosynthesize food that supports thriving populations of reef dwellers in nutrient-poor waters.

Nearly every group of organisms has members that have formed close partnerships for feeding, cleaning, or protection. The physiology and patterns of inheritance of modern symbionts provide analogies for evaluating the hypothesis that cell organelles arose through symbiosis.

What kind of world did the early prokaryotes inhabit? Were the conditions of the planet and its atmosphere those we know today? The Earth's surface, oceans, and atmosphere have been so profoundly altered by the activities of living forms on the planet that to answer these questions, one must turn to studies of our lifeless neighboring planets. The Earth condensed out of a cloud of dust and gases that formed it and the other planets of our solar system. Astrophysicists postulate that most of the major bodies of the solar system originated during the same period, about five billion years ago. Photographs taken from orbiting spacecraft show similarly cratered surfaces on the Moon, Mercury, Venus, and Mars and its moons. The fact that the oldest rocks taken from the Moon's surface, as well as meteorites found on Earth, are all about 4.5 billion years old also supports the idea of a common origin of the major bodies of the solar system.

One can, therefore, consider Mars and Venus as sterile Earth-like places with similar planetary histories and make some good guesses about how life has modified the surface of our planet. (The results of the Russian Venera 9 and 10 probes of Venus, as well as the United States' Viking mission to Mars in 1976, also suggest that this assumption is plausible.) One of the most conspicuous differences between the Earth and its neighbors is the large concentration of oxygen found in the Earth's atmosphere. The atmospheres of both Venus and Mars contain about 98 percent carbon dioxide and far less than 1 percent oxygen (they also have about 2 percent nitrogen and some water vapor), whereas the Earth today has nearly 21 percent oxygen and only 0.03 percent carbon dioxide (and 79 percent nitrogen). When the Earth first formed, its atmosphere probably resembled the atmospheres of its neighboring planets at that time.

Biological considerations also support the theory that the young Earth's atmosphere contained no free oxygen. Life originated on the Earth through the formation and interaction of prebiotic compounds: nonbiologically produced amino acids, nucleotides, and sugars. Such chemical compounds simply do not accumulate in the presence of oxygen, which reacts with them and destroys them as soon as they form. The first cells on Earth, then, must have arisen in the absence of oxygen.

Primitive bacteria—those believed to be most directly descended from our earliest ancestor cells—are poisoned by oxygen. They have no chemical or other means of protection against the gas, and their cell material burns up, in effect, if exposed to it. Such bacteria, called obligate anaerobes, live by fermentation, taking up organic compounds and generating ATP anaerobically. It is reasonable to assume that they evolved in the absence of oxygen.

In time, the supply of organic compounds became

limited; the evolution of photosynthetic apparatus, which enabled cells to manufacture the organic compounds they needed from inorganic compounds using light as energy, occurred. The first photosynthesizers, however, were also anaerobic bacteria; none of the earliest forms of photosynthesis generated oxygen.

How, then, did the Earth's atmosphere become oxygenic—a transition totally unpredictable from the laws of chemistry and physics? And when did the transition occur? To answer these questions, we have to look at the organisms that succeeded the earliest photosynthetic bacteria. These successors were the blue-green algae, a misnamed group of photosynthesizers that are not algae, nor are they always blue-green in color. Today, in recognition of the essential affinity of these microbes with other bacteria, biologists use the term *blue-green bacteria*, or *cyanobacteria*. These cyanobacteria were probably the first organisms to give off oxygen as a waste product of their photosynthesis.

There is direct fossil evidence for the first appearance of cyanobacteria as early as 3.5 billion years ago, and a great proliferation and diversification of cyanobacteria about 2.5 billion years ago, a date that fits nicely with evidence from the geological record, which shows two-billion-year-old rocks containing oxidized forms of minerals. After the oxygen-loving minerals (like iron) reacted with bacterial oxygen, the upsurge of this reactive gas in the Earth's atmosphere must have been due to the worldwide proliferation of these bacteria. Never before or since have organisms on Earth so profoundly affected the atmosphere.

The rise of aerobic photosynthesis was a global catastrophe. Because oxygen was toxic to early life, it became an increasingly serious pollutant. Like automobile waste products, this pollutant even threatened the producers themselves, the cyanobacteria. The resolution of the oxygen crisis was a turning point in the history of the cell: microbes evolved the capacity to use in respiration the oxygen that they produced. This solution not only protected them; it also provided them with additional energy, because oxygen respiration generates far more ATP than fermentation does. In time, as the concentration of atmospheric oxygen rose, cells of many nonphotosynthetic species evolved that required oxygen for their metabolic processes; these were the first obligate aerobes. They put the potentially poisonous oxygen to use in the elegant innovation of aerobic respiration. By this means, cells could generate enough ATP to grow larger and perform more sophisticated functions. About 600 million years ago, at the beginning of the Cambrian geological period, there was a virtual explosion of large forms of animal and photosynthetic life, their visible success the result of the miniaturized achievements of their microscopic ancestors.

For generations, the Cambrian rocks were thought to be the beginning of the fossil record. The time before the Cambrian is still often listed as a vast undifferentiated era, the "Precambrian," on geological time charts. Now enough is known about those times to recognize divisions in the Precambrian: the Hadean, the Archean, and the Proterozoic Aeons. The Hadean, whose name derives from Hades, the hot and chaotic underworld of Greek mythology, extended from 4.6 billion to 3.8 billion years ago. During this time, the Earth and its Moon took form as solid bodies. Meteorites and lunar rocks date from this period, but so much debris hit the Earth and there was such shuffling and melting of material on its surface that no terrestrial rocks remain from the Hadean. The Archean Aeon, extending from about 3.8 billion to 2.6 billion years ago, saw the formation of the Earth's long-lived crustal features, the appearance of life on the planet, and development of the major metabolic strategies, including fermentation, photosynthesis, and the ability to convert atmospheric nitrogen to a form usable by cells. The beginning of the Proterozoic Aeon, about 2.6 billion years ago, is marked by a change in the character of the surface rocks. This aeon extended to the beginning of the Cambrian Period, about 600 million years ago. During the Proterozoic, eukaryotic cells developed two-parent sexual reproduction, giving rise to animal and plant ancestors. Eukaryotes of many forms evolved; by about a billion years ago, large algae several centimeters in diameter had evolved; and near the very end of the aeon, about 700 million years ago, the first soft-bodied multicellular animals appeared. In keeping with this scale of time divisions, the time represented by the "classical" fossil record—trilobites, the first land plants and animals, the vast forests whose remains form our coal beds, the dinosaurs and wooly mammoths—has been named the Phanerozoic Aeon.

The more that is learned about the Earth, the clearer it is that our planet's surface has been highly altered by the origin, evolution, and growth of life on it. As life expands, it alters the composition, temperature, and chemical nature of the atmosphere and the composition, texture, and diversity of the Earth's surface. The surface environment and the organisms on it have been evolving together for billions of years. My narrative has traced the evolution of cells that became structurally and functionally more intricate and that gave rise to many groups of larger and more elaborate organisms, but it would be a misreading of the evolutionary record to think of these events as a kind of upward progression.

Some authors have claimed that evolution has been "progressive," leading to "higher" and therefore better life forms. One must realize that even three billion years ago, neatly functioning atmospheric cycles were modulated by organisms. By two billion years ago the cyanobacteria had made drastic changes in the atmosphere. It is doubtful that any organisms since then have had such a profound effect on the planet. If the vast stretch of pre-Phanerozoic time once seemed uneventful, it was because we lacked the tools to examine it. We now realize that it was the age of prokaryotic microbes. Without their achievements—their adaptations to extreme environments, their exchanges with the atmosphere, and their production of oxygen—the spectacular spread of eukaryotes would never have been possible. Without the prokaryotes' continuing activities, neither we nor the animals and plants on which we directly depend would continue to exist.

We consider naive the early Darwinian view of "nature red in tooth and claw." Now we see ourselves as products of cellular interaction.The eukaryotic cell is built up from other cells; it is a community of interacting microbes. Partnerships between cells once foreign and even enemies to each other are at the very roots of our being. They are the basis of the continually outward expansion of life on Earth.

6

HENRI ATLAN

Uncommon Finalities

ONE OF THE main problems in biology for a long time has been the problem of final causes. Their existence was obvious for Aristotle without any reference to theology. However this question became a real problem in the last century when scientists did not want to accept the idea of final causes as a part of scientific explanation. I do not want to go into the reason they did not want to do so now, since I shall come back to this point, but in biology especially this problem was crucial, because plain observations of living systems impose the idea of a finalism of some sort: we know in advance what will happen to an egg, and everything seems to happen as if the development of the egg toward the adult form is determined by the final state as much as, or even more than, by the previous state. The question, of course, was always how to explain that, and this was the subject of the great classical debate between the vitalists and the mechanists. The former believed that the only way to explain this kind of phenomenon was to assume vital forces directing the evolution of the living system toward its final stage, while the latter did not want to accept such an explanation, and preferred to look for physicochemical causal phenomena which would explain the process as a sequence of causes and effects.

Now, when we look into the other sciences, such as physics, for example, we find several instances in which we are dealing with something that looks like final causes also. There, this does not trigger a scandal, as is the case in biology. For example, whenever we have a physical law expressed by an *extremum* principle, such as a law of minimization of free energy, or maximization of entropy, or a law of minimization of potential, (i.e., whenever we have a physical phenomenon whose evolution in time is expressed by a mathematical law stating that some quantity must reach maximum or minimum value), then in fact we are dealing with a kind of finalistic explanation: phenomena are described by pointing to a final state, defined by the *extremum* value, to predict their temporal evolution.

Why is it that in some instances no one is bothered by such laws, which are finalistic to some extent, and in other instances, such as living systems, we do not want to hear about finalistic explanations? It is as if in the natural sciences there are two kinds of finalism, one which is a good finalism, and one which is a bad one. Now, what makes the finalism good or bad? I think that what makes it good or bad, at least for scientific thought, is whether we are talking about a conscious finalism or a nonconscious one. The reason physicists accept explanations based on *extremum* principles is that these principles, first of all, are described within a fairly well-established, precise, and explicit mathematical formalism; second, they do not assume a conscious will to orient the process. On the other hand, biologists reject the idea of final causes to explain the temporal evolution of living systems because what is implied always, behind vital forces or whatever, is something like a conscious will responsible for the orientation of the temporal process.

Now to make a long story short, it was to solve this kind of problem in biology that the idea of the genetic program was invented. This concept has had a tremendous impact in the development of modern biology, although its explanatory value is very weak, as I will try to show. The concept of the genetic program was supposed to solve the problem of conscious versus unconscious finality, and the main reason it is so weak is precisely because it has not succeeded in doing what it was supposed to do. Littendrigh was the first, in the fifties, to use the word "teleonomy" to replace teleology, knowing perfectly well that there was no difference in the literal meanings of the words; he was proposing a new word just to stress the difference between what he stated as an "end-seeking machine" or "end-seeking process" and a purposeful one. Of course the "good" finality was supposed to be that of an end-seeking process, while the "bad" finality was that of a purposeful one. Teleonomy was supposed to deal with trying to describe an end-seeking process, and was proposed as a scientific concept instead of teleology, which, traditionally, had to do with purpose.

Now, a very few years after that, the idea of the computer program was proposed, and, miraculously, it looked perfectly fit to give an operational content to the concept of teleonomy, for two reasons. First, the computer, a deterministic machine directed by the execution of a program, appeared to be a realization of an end-seeking machine, deterministic but end-seeking, especially if you forget about the programmer. Second, the discovery of the genetic code, which occurred at about the same time, gave the impression that in living systems we were dealing not only with a metaphor but with a real thing, since the physical support for the genetic information was found to be in identified molecules. Therefore, the idea of using the language of computer sciences to describe processes in living systems had advantages. But, as I said before, although the concept of the genetic program has had an important operational value in triggering new experiments and discoveries, it has a weak explanatory value, precisely because it still has some conscious finality in it.

Why is it so important to get rid of conscious finality in the analysis of a living organism? One classical reason is a kind of economy. If we assume a consciousness at work in nonhuman living organisms, as well as a kind of vital principle, since this is something of which we have no experience, it is of necessity an ad hoc hypothesis and does not add very much to the understanding of the system; moreover, it prevents us from doing deeper research into the physical chemistry of the living organism. Therefore, from a methodological point of view it is a bad hypothesis. Another reason seems to me at least as important if not more. If we assume a consciousness at work in the organization of living systems, then the determination of the end appears akin to the determination of the future in a conscious planning. This implies the impossibility of newness, the impossibility of unexpectedness, and, in the end, it implies the negation of time, because the future is determined by a conscious plan which itself is nothing other than a projection of the past, even if there is some modification. Therefore time cannot bring with it some radically new or radically unforeseeable thing. That is why assuming a consciousness at work in living systems, and even to a broader extent in nature generally, nonhuman nature, has very important disadvantages.

One of the Jewish masters of the eighteenth century expressed this kind of idea in a very interesting way. He was trying to put together (in couples) three different levels of the living "soul" with the three tenses of time, past, present, and future. The soul was understood not so much as a physical force but as what makes the organism alive, with different levels, of which he was describing three. One had to do with the unconscious life of the body, another with the feelings or sensations, and the third with the intellect. So the Jewish master tried to pair off these three levels or aspects with the three tenses of time, based on the idea that time and living souls are a kind of couple that makes the world alive. And the question was, which tense went with which aspect? Now the almost self-evident association that would be made by most people would be that the future would be associated with intellect, but this is precisely what he did not want. On the contrary, he stated that the future is associated with the unconscious part of the body, an unconscious part that makes the body alive, whereas the intellect has to do only with the past, and the feelings with the present. The argument for this was simply that the future is unknown and therefore cannot be associated with the intellect, which has to do with knowledge; what is known must concern only the past, and what is unknown has to be associated with the unconscious.

The need for the possibility of an unknown future, bringing real newness with it, is the main reason why, in my opinion, consciousness in nature is a bad hypothesis. That is why even though the concept of teleonomy in the form of the genetic program as it was developed in biology is not a perfect one, I think it should not be rejected. We should not go back to a kind of vitalistic approach because at least teleonomy, with all its pitfalls, is the beginning of the representation of a finality without consciousness and without intention. In other words, I think that we must try to improve the concept while keeping its two advantages: namely, that first it tries to avoid the additional ad hoc hypothesis of a conscious finality, a kind of cosmic consciousness at work in phylogenesis and ontogenesis; second, and maybe more importantly as I have tried to explain, it gives us the possibility to account for our experience of newness, of a real creative activity capable of producing newness only inasmuch as it is unconscious.

However, the weakness in the concept of the genetic program is that it can mislead by giving the idea that it works essentially like a real computer program. Unfortunately many biologists and philosophers of science who have taken the metaphor of the genetic program literally have been put back, without knowing it, into the classical vitalistic attitude; the only difference is that in their language God and the vital force have been replaced by natural selection. The reason for this is that the metaphor of the program, if you take it literally, leads you to ask the question of the origin of the finality, which amounts to asking: what is the programmer? And the classical answer to this question is: the programmer is natural selection. But the problem is that we do not have the slightest idea how natural selection can write a computer program. We don't know the language, and the fact that we know the genetic code has nothing to do with a computer programming language; it is, at best, a lexicon. Only if we knew how a given genotype can express itself as a definite phenotype in a given environment would we know approximately the computer language of the genetic program. In other words, only if we knew all the mechanisms of genetic regulation and genetic expression for a phenotype in a given environment would we have, maybe, some insight into the language of the genetic program. Otherwise we do not know at all how natural selection can write something which would look like a computer program. That is why the term natural selection as it is now used is a type of magical invocation, a magical word to be used every time one has to explain a given adapted and finalized natural organization. This is another aspect of how, if we are too stuck to the idea of programming as an explanation for adaptation, then unavoidably we are led to the idea of a kind of homunculus acting out there with some intent in order to do the best possible thing in the world.

These are the problems which are the background for the research, formal and otherwise, about self-organization. In other words, it is because we are not completely satisfied with the idea of the genetic program that we want to look at how it would be possible to conceive other mechanisms for self-organization. To pursue this idea of the program further would lead to a semi-science-fiction idea of a self-programming program.

Now, there is something special in the idea of self-organization which has been seen already by Ashby and Heinz Von Foerster, who were the pioneers in this field, and it is that self-organization in an absolute sense cannot exist. In other words, if you look at the organization of a system as the set of rules which makes the system work, and if you think about a change in the rules, then the organization is not only the set of rules but also what directs, what makes this set of rules work; and the idea of a *self*-organizing system would be one which would be able to *change* its set of rules in such a way that, for example, the new organization would be adapted to another situation or would do something else. But of course if the system changes its set of rules, then we would be looking for: what are the rules which govern the change of rules? And it is these rules which govern the change of the rules that would be called the organization of the system and would not be changing. Then if the system would be able to change by itself the rules which change the rules of the system, again we would look for the laws of the change of the rules of the change of the rules. And we would always end up with something which would not change. Therefore, a purely self-organizing system cannot exist; something that changes the rules must come from the outside, and, if this is so, there is no reason to talk about self-organization. However, two different things can come from the outside. One is a program, that is, a set of rules telling the system how it must change its set of rules—and then, of course, there is no reason to call the system self-organizing. The second possibility is that what comes from the outside is not a program but just random perturbations. Random perturbations, in general, have been considered as something which is not good for an organized system; they were thought of as only being able to disorganize a system, to cause more disorder in it. Now, if one can think of a system in which random perturbations result not only in disorganization, but also produce a change in the organization of the system, such that the system not only continues to function but does so in a different way (perhaps, for example, a way more appropriate to a different environment), then such a system could be called self-organizing, although not in a strict sense.

The question now becomes: how can such things exist? I do not want to get too technical, but this has been described by Von Foerster, who called it "Order from Noise." I extended the idea a little bit and called it "Complexity from Noise," because I think this is simpler to understand. The main point of this work is as follows: Let us take a network of communicating elements. Random perturbations (called noise in communication theory) will untie the constraints within this network by creating some ambiguity in the communication. However, this detrimental effect, the fact that some ambiguity is created, on the one hand decreases, of course, the information transmitted from one element to another. But on the other hand this same effect, the same ambiguity, can be counted as positive and may be viewed at a different, more integrated level of organization. For this higher level of organization, untightening the constraints may result in a different organization with more diversity, and this new organization may have different and stronger adaptive properties. (Of course, this can be true only up to a certain point, and providing that the system is redundant enough so that it will continue to function.) This means that when you go from one level to the other, you have a change in sign from negative to positive, and this appears in formal studies as a change in the algebraic sign of a quanitity called ambiguity, which, while negative at the elementary level, appears as positive at a more integrated level. In other words, you start by looking at the different elementary parts interconnected with one another; then you shake them randomly and destroy bonds between them, thereby creating more disorder. But this additional disorder can be viewed (up to a certain point, if the system continues to function) as more complexity, meaning that the connections between the elements have been reorganized and are seen now at the more integrated level as forming a new organization with fewer connections. This condition of having fewer connections would mean only more disorder if the new organization does not work, but it means complexity if it works in some way.

Now this change of sign seems to be only a particular case of something more general, and more trivial to some extent, which has to do with what happens when one goes from an elementary level of organization to a more integrated level. Whenever we go from one level to the other, we encounter a kind of transformation in relationships between elements which are viewed as separated entities at one level and unified ones at the other. If you look, for example, at how atoms are put together to make molecules, you find that what makes the atoms different from one another and helps to separate them is precisely what makes them link together to produce molecules. The same holds true if you look at molecules associated to produce supramolecular structures or cellular structures. If you want to distinguish between elements, then you look at them one at a time and try to identify some properties which make them different from one another. But when the same elements are put together to produce a more integrated unit, then these very same features which help distinguish between them become the origin of the bond uniting them, and are what makes the more integrated unit. So whenever we go from one level to another it seems that we are forced to change our point of view from looking at the features of the individual elements in a way which helps us distinguish between them, to looking at the same features, but now at how they are put together in order to create the new level of organization.

I will come back to this later, but what I want to note now is that it is here, at this critical point where we are dealing with the articulation of one level with respect to the other, that we encounter also a puzzling problem, which can be stated, in terms of information theory, as the creation of meaning. Again, to make things short, I want to present this in a very schematic way. If a system is going to function, then there must be exchange of information, of course with the surroundings, but also within the system itself; and, within the system, there must be exchange of information not only between constitutive parts, but also between levels of organization. However, in most cases we do not have access to this latter transmission of information. We do not know how one level communicates with another, and we don't know that for a very simple reason: because it is *we* who are creating the different levels by means of different techniques of observation and experimentation. Therefore, what happens between these levels is absolutely obscure to us since we have no access to it.

What the idea of "Complexity from Noise" expresses, essentially, is that in this situation, in which we are as observers of a system with different levels, the untightening of the constraints at one level will create functional complexity rather than mere organization, under the condition that the complexity will be felt at a different, more integrated level. In other words, this idea implies that the new relations created by the untightening will be integrated into a new organization, with more diversity and less redundancy, but only under the condition that there will be a functional way to do that. (That is why this theory is a theory of necessary conditions, and not sufficient conditions, because it is a theory based on the lack of knowledge that we have about *how* the system manages to organize itself. For example, among other things, the high initial redundancy which is a necessary condition for self-organization is certainly not a sufficient one; in addition the system must be able to make use of the new organization, i.e., the system must be able to reorganize the new state of connections in a functional way.) Thus, self-organization, viewed as noise-induced disorganization followed by reorganization cannot exist without interplays between different levels. And describing self-organization as the utilization of noise to create functional complexity in fact amounts to describing the creation of new meaning, still unknown to the observer, but new meaning in the information transmitted from one level to another. However, the description is made in a negative way because, when done within the formalism of information theory, it makes use of a formalism from which the meaning of information is explicitly absent, although its existence is always implicit in the actual functioning of an organized system. In other words, what appears to the external observer as organizational randomness, *le hasard organisationel,* implies the creation of new meaning, as yet unknown to the observer, between levels within the system. Thus, it seems to me that this problem of the articulation of one level with respect to another constitutes the frontiers of our scientific knowledge.

Now let me come back to a more general change, the one from separation to reunion, because I think this may reveal the most general aspect of system organization. When you are interested in the individual elements, you try to distinguish between them, and therefore you stress how they are different from one another, but when you are interested in the way all these elements are combined to produce an integrated unit then you are forced to look at what they have in common; and it so happens that the two are the same. What makes the elements different is precisely what they are forced to put in common to constitute a more integrated unit. To some extent this is a trivial thing that amounts to just a change in the point of view. However, what is less trivial is the relationship between the transformation from separation to reunion which takes place between two levels and the emergence of new properties in the more general level as compared with the more elementary one. And you find such a relationship everywhere. You find it when you go from atoms to molecules, from molecules to cells, from cells to organisms, and so on. For example, when you go from atoms to molecules new properties of matter are revealed, the existence of chemical affinities, or, more generally, the chemical properties of matter, which are new compared to the physical properties of atoms. The same thing happens in going from molecules to cells: something new emerges, the cybernetic and organizational properties of cellular organization, or the biological properties of cells, which are new compared to the chemical properties of molecules. The same thing happens again when you move from cells to organisms and discover the physiological and differential properties of the organism, which are new compared to the cellular properties; and so forth . . . the pyschological and behavioral properties of animal behavior and human mind are new compared to the physiological properties of the brain.

By this point you may have noticed that in fact the emergence of new specific properties at a given level implies the existence of a new scientific discipline: physics, chemistry, cell biology, cell physiology, psychology, sociology, geology, "gaiology," etc. And a very important question is: to what extent are the different levels of organization real and to what extent are they due to our operation of cutting reality into different levels due to different techniques of observation? It is obvious that the role of this operation is very important, as evidenced by our image of the living cell. Surely no one has ever seen a living cell the way we represent it to ourselves, because this representation is a mental reconstruction and superposition of all kinds of properties which are observed at very different levels. What we know about living cells comes from biochemistry, but biochemistry implies the destruction of the cell; or it comes from different techniques of microscopy, some of which do not destroy the living cell and some of which do. Every one of these techniques reveals bits of information about what a living cell is, but it is impossible to observe the different levels of organization *together with the same accuracy.* If you want to observe the cell at the level of its molecular structure, you have to use biochemistry and then you destroy the cell. If you want to observe it by electron microscopy, then you lose a lot of information about the biochemistry, and so on.

It is as if every technique of observation is able to focus only on one level and not the others, and it is our mental reconstruction which puts them together. Then, the question is how articulations between levels can be represented, since by construction we cannot have access to them. This is a question which cannot have an answer, inasmuch as the levels of organization are created by our means of observation, and the articulations between them, therefore, are something which we cannot observe. However, it is probably at this place that, apparently, the source of the autonomy of a living system lies. We have seen what can be said indirectly about it: it is a place where a change of sign seems to take place, or a transformation between separation and reunion; also it is the place of the creation of new meaning characteristic of self-organization; and so on. But all these things are only formal descriptions and, in a sense, are only different ways of saying that we do not have direct access to this place. Now, it is interesting to observe that the whole situation can be changed overnight if new techniques are discovered which give access to this thing which was viewed as a place of articulation between one level and another. We have been provided with a spectacular example of such a change by what happened with the advent of molecular biology. There was a long period when we had to deal with two different levels of organization, which were at the same time two different disciplines, chemistry and biology in this case. These two disciplines had nothing in common as far as the techniques were concerned, as far as the language of their theories was concerned, and therefore it was impossible to understand how one level could be responsible for the other or could be articulated from the other. The gap between chemistry and cell biology was what enlivened theoretical problems in biology for a long time, until some techniques were discovered to analyze the structure of macromolecules and to understand how they duplicate themselves, how they produce polymerizations in their specific order as occurs in living cells, and so on. As a result, suddenly the gap between chemistry and cell biology seemed to have disappeared, as the newly discovered techniques were applied to accomplish just that. In fact, this was not completely true. The gap did not disappear, because as soon as the new techniques were found a *new discipline* was created—which is called molecular biology—and it has developed its own techniques of observation, its own theoretical tools, its own language. And the question still remains, or rather new questions are raised, about how this field, the level of molecular biology, is going to be related to the level of chemistry on the one hand, or to cell biology on the other. It seems as if we are playing an endless game: either there is a level of articulation between levels to which we have no access, or, if we discover an access to it, in so doing we just create a new level and at the same time two new articulations to which we have no access. Nevertheless, one can say that there is something to be gained, because every time this happens the gap is made narrower. It can be seen as an asymptotical process, which will end only at infinity, of filling the gaps partially and narrowing them, while new gaps are made between new disciplines.

Another example of a new discipline partially filling the gap of articulation between two levels of organization, and creating two new—and, we hope, narrower—gaps with new questions is that of language. However, the way we look at language from this point of view has several interesting features. First of all, language can be viewed as located between two different levels of organization, namely between body and thought, or between body and mind, if you want, which means, in a more positivistic way, between physics and psychology. In other words, whether we want to consider the levels of organization as ontological levels or to consider them just as the results of disciplines, language would appear as intermediate between body and mind or between physics and psychology. Then, as was the case with molecular biology, language has become by itself a level of organization and the object of a scientific discipline. In addition, and probably the most interesting thing, it is within language that all the other levels are described. And finally, in a recursive way, language is in itself a multilevel, self-organizing system, where we find, again, the problem of the articulation between different levels. Within language, one can describe several levels of organization: the semantic level of the words (i.e., their meaning), the syntactic level of the sentences (i.e., the rules by which the words are combined to produce sentences), the semantic aspect of sentences, and so on. Also, if you look in the other direction, you find a sequence of different levels within the word itself. If you consider the level where different signs are associated to produce a word, you find a kind of syntax; in letters or signs you find also a kind of semantics, although very elementary, in which the letters or signs are viewed with their individual meanings, as is the case with ideograms. The interesting thing is that whenever we go from one level to the other, we are forced to switch from one aspect to the other. If we look at the individual letters, for example, as signs with a kind of ideographic meaning, as is the case in Hebrew, then as soon as these signs are combined to produce words this ideographic meaning is forgotten, and new entities are created at a new level, that of words. It is the same when we go from words to sentences, from sentences to utterances, and so on. Now, if we ask how one level is articulated with respect to another, every time we encounter the same kind of phenomenon that I have talked about before, that is, a transformation from separation into reunion. What makes different letters separate from one another is precisely what makes it possible for them to be put together in a word. Similarly, it is what causes separations between words that puts the words together to produce sentences. So, in the end, it is in this interplay between different levels that the meaning of the whole seems to take place. And if we ask ourselves where we could locate the meaning of language, then we end up with the strange but suggestive idea that it is to be found in the blanks between the writing, or in the space between the letters. This idea, which I have found in a Cabalist writing of the beginning of the century, although not exactly in the context I put it here, is a seemingly paradoxical way to suggest that the meaning cannot be localized somewhere as something static. Rather it is created as the result of a process of self-organization of language, the articulation between levels being the critical parts in this process.

To summarize, the frontiers of knowledge are to be found not only in the infinitely large, as usually thought, but in the articulations between various levels of organization of reality. It is important to realize that these different levels correspond to different fields of knowledge. Their techniques and discourses are separated, and they do not touch one another at their limits where they are articulated from one another. We have only very limited means to talk about these articulations, because they appear in between the different fields of scientific knowledge; and, therefore, we cannot have a direct access to them. Nevertheless, it is at this place, which looks like a nowhere, that the origin of the autonomy of a complex system is to be found. As we have seen, this is also where the creation of meaning, with its aspect of self-reference, seems to take place. That is why, probably, that what we find at these frontiers, as a kind of shadow, is the question of the recursiveness of the self.

However, at this point we must be very careful not to be trapped into a spiritualistic mistake which would make us act as if we already knew what the self is, starting from our subjective experience of self-consciousness. The self I am talking about is not necessarily human, and, therefore, to a very large extent it is unconscious; similarly, we are trying to understand, objectively, so to speak, what meaning is in nonhuman communication systems, so that our subjective experience of meaning and self appears as a particular case of something more general. A good example of such a nonhuman self is given by the molecular and cellular self of the immune system, or by a computer program which would be able to program itself. In this approach, what characterizes our subjectivity as a particular case is the special position of its observer both inside and outside the system at the same time. There is no circularity in taking into account the role of the observer to characterize subjectivity because the observer I am talking about is *not* our subjectivity: it is the so-called ideal physical observer, that is, a kind of idealized set of operations, of measurement, and of combinations of these operations by logical relationships. Thus, we are dealing here not with a logical circularity, but with a phenomenon of recursivity in which a fundamental role is played by language with its three aspects as we have seen them: language (spoken and written) is found at the articulation between two levels of organization, the mind and the brain; at the same time, it is the tool by which these and all levels are described and analyzed; and, at the same time again, it is itself a whole self-organizing system with several levels.

As a last word, I want to add that stressing the aspect of unconscious finalities in our experience of the world does not mean that I am advocating a philosophy of "blind will." I think that our consciousness is efficient in projecting itself and acting on reality. In other words, our human reason and human mind are efficient indeed in coping with reality, and it would be silly to despise them. However, they are efficient in a limited way, and I think that the limitations of our capacity of analysis—which are found at the articulations between different levels of self-organization—are fortunate. These limitations are fortunate because they maintain the possibility of the emergence of newness, of the unpredicted. It is at this junction, which makes time creative, that we see appear, as a *shadow*, what we could call a self-*unconsciousness*, not necessarily human. And we can reach this junction by three different paths: one which explores the organizational role of randomness; another which explores the role of the creation of meaning; and a third, that of the autonomy of the self, which brings together "the knower, the known, and knowledge."

PART TWO

Gaia Politique

7

JOHN TODD

An Ecological Economic Order

NOT LONG BEFORE his death I was with E. F. Schumacher at an appropriate technology congress on the Indonesian island of Bali. Although I had known Fritz for years, my most cherished memory of him is from the Balinese countryside. We were visiting an international development project that included a modern fish culture facility. Unlike the rest of the food culture on the island, the demonstration fish farm seemed alien, with its fences, rectangular ponds, and separation from the agriculture and the villages. Like a prison in our society, it was separate from the ordinary interwoven fabric of Balinese culture.

Later that day we visited a temple. The water, trees, architecture, and gardens expressed a deep harmony and what seemed to me a merger between mind, nature, and the sacred. As the sun fell, Fritz spoke of how trees were the most powerful of transformative tools and their planting and tending a fundamental act. For him trees were the starting point for creating social and biological equity among peoples and regions of the earth.

Our conversation together inspired some ideas which follow. I owe E. F. Schumacher a debt for helping me see economics as if people and nature mattered. Subsequently I have come to believe that a new sustainable economic order can be established with ecologically based enterprises. Further, the conceptual bases of these enterprises are similar when applied in rich industrial nations or in poorer tropical countries. If this thesis is correct, applied ecology has the intrinsic potential to dissolve old divisions between north and south, industrial and agrarian, and rich and poor. This is so because ecological knowledge can be applied universally and, equally importantly, it can often be directly substituted for capital and for non-renewable resources. In the same sense that Fritz Schumacher spoke of trees, it has the ability to increase equity on a global scale.

Ecology as the basis for design is the framework of this new economic order. It needs to be combined with a view in which the earth is seen as a sentient being, a Gaian world view, and our obligations as humans are not just to ourselves, but to all of life. Earth stewardship then becomes the larger framework within which ecological design and technologies exist. One day it may be possible for political and social systems to mirror the broad workings of nature, and current divisions of left vs. right, centralist vs. decentralist, expansionist vs. steady state, bioregional vs. the nation-state will be transformed into a systemic Gaian world organization and order.

But change, even on a Gaian scale, has to begin with small, tangible, and concrete steps. When I first began working with ecological concepts that might serve humanity at the New Alchemy Institute, my associates and I started with a question, "Can nature be the basis of design and are there ecological models to prove this?"

We started with food and agreed that the contemporary mechanistic agricultural model would in the long run fail to feed the planet. We looked for other models to guide us. The larger workings of nature provided us with clues. We sought out several places where nature is extremely bountiful and made a shopping list of the attributes particularly unique to those places. As patterns gradually emerged, this effort proved directly fruitful. We also sought out places which, under the guiding hand of humans, have been bountiful for millennia. This was significant because humans normally destroy their biological capital. We wanted to learn what stable cultures know about caring for their lands.

A farm near Banding in central Java was rich with clues. It had maintained and possibly increased its fertility over centuries. The farm was located on a hillside that was particularly vulnerable to erosion which was prevented by mimicking nature's most efficient erosion control strategy, namely, tree-covered slopes. It was not a wild forest, but a domestic one in which the biota were fruit, nut, fuel, and fodder trees useful to humans. Nevertheless, it had some of the structural integrity found in the wild. Without the trees on the hills it would have been very hard to sustain the land's fertility. The farm received its water from an aqueduct flowing across the slope halfway up. The water came from a farm higher up and arrived in a clean, relatively pure state. Upon entering the farm it was, within a short distance, intentionally polluted first by passing it directly under slatted livestock barns and then under the household latrine.

Although it might appear shocking at first glance, the livestock and household sewage was then utilized in a clever way. The solids were "digested" by a few fish whose sole function was to provide primary waste treatment. The nutrient-laden sewage was then aerated and exposed to light by passing over a low waterfall. Secondary and tertiary treatment were agricultural. The sewage was used to irrigate and fertilize vegetable crops planted in raised beds. The nutrient-rich water flowed down the channels and dispersed laterally into the soil to feed the crop roots. It is important to note that the secondary sewage was not applied directly to the crops but to the soil. The water emerged from the raised bed crop garden with nutrients removed and at least in an equivalent condition to our tertiary treatment. It then flowed into a system that requires pure water, namely, a small hatchery for baby fish. Here in the hatchery pond the young fish began the enrichment cycle again by slightly fertilizing the water with their wastes. This triggered the growth of algae and microscopic animals that helped feed the young fish. This biota was also carried along with the current to add nutrients and feeds to the larger fish cultured in grow-out ponds below. These highly enriched grow-out ponds fertilized the rice paddies that were just downstream. The rapidly growing rice used up the nutrients and purified the water before releasing it again to a community pond in the basin below.

The intriguing thing about the farm was that it was a complete agricultural microcosm. There was a balance not seen in Western farming. The trees, soils, vegetable crops, livestock, water, and fish were all linked to create a whole symbiotic system in which no one element was allowed to dominate. Such a system, while beautifully efficient and productive, can also be vulnerable to abuse. One single toxin, like a pesticide, will kill the fish and unravel the system. The lesson in it is that we can create ecological agri-systems and let nature do the recycling, or we can manage a complex system chemically and ultimately destroy its underlying structure. At New Alchemy, when we began to design food growing ecosystems, we tried to keep the biological relationships of the Java farm intact. Even years later the design of world voyaging ocean arks paid homage to those generations of Javanese farmers.

There are lessons from all over the world, even from endangered places. As Shakespeare said, ". . . sermons in stones, books in the running brooks, and good in everything." Soils are dying around the world. Deforestation, overgrazing, fire, and erosion are the primary villains. To understand the importance of soils, and how we are enjoined with them, we need to realize that soils are alive, meta-organisms comprised of myriads of different kinds of living creatures. When they are exposed to sunlight, blowing winds, and mineralization, they become increasingly lifeless and porous, losing their ability to retain rainwater near the surface. Most of the world's spreading deserts follow in the wake of soils becoming more porous and devoid of rich microscopic life.

Perhaps one of the single greatest challenges facing humanity is the restoration and re-creation of soils. They need to be given back their organic matter, humus, and moisture-retaining qualities. Without healthy soils human economies cannot be sustained for long.

Several years ago we visited an atoll in the Seychelles in the middle of the Indian Ocean. The soils of coral islands don't hold water, as they are notoriously porous. Rainwater percolates quickly through the soil and collects in underground lenses. On the atoll we visited, the one hundred villagers had almost pumped their fresh water lens dry, and salt water was beginning to intrude and contaminate their water. Within a few years the inhabitants would have to abandon their islands.

Their seemingly intractable problem could be solved if somehow impermeable basins could be created to capture rainwater during the monsoons. But the soils were too porous to consider a surface pond idea as a viable option. However, I remembered the research of two biologists who had discovered a strange anomaly in Russia. They found that on top of hills comprised of rubble mounds, ponds or small lakes would occasionally be found. Since the underlying soils were incapable of holding rainwater, there had to be some mechanism that sealed these ponds so that they could capture and hold rain. They then discovered a comparatively rare process in which microorganisms, in concert with organic matter, combine to produce a biological sealant. This sealant formed a liner in the natural basins which then held water. They called the process gley formation.

On the atoll we decided to mimic the process discovered by the Russians in the very different tropical coral island environment. We hoped, if the conditions were just right, gley formation might take place quickly. The challenge was getting them right. We dug a small lake with a backhoe. We found the husks of coconuts to contain the necessary carbon and fiber component, which we shredded and placed in a six-inch layer over the bottom and sides. For our source of nitrogen we collected the ubiquitous wild papaya and chopped up their stems, branches, and fruits. They were placed in a six-inch layer above the husks. Finally six inches of sand were placed on top of the husks and papaya. The Russians had found that gley forms in the absence of oxygen so the basin was packed down to drive out oxygen. A small amount of water from the lens was pumped in to flood the bottom. To our great pleasure, when the monsoon rains came shortly thereafter, the basin filled with water and it stayed.

The pond today is a source of irrigation water, the home to cultured fish, and a haven for wild fowl including migratory birds. The experiment opens up a whole range of ecological and economic possibilities. Not only can coral islands be ecologically and socially diversified, the very same process can be used wherever there is a need to store seasonal rains. I can foresee, throughout the world, previously barren landscapes nurtured by small gley-created impoundments which are the epicenters for restoration of damaged environments.

The new freshwater source inspired an experiment to make the atoll's alkaline and nutrient-poor soils capable of growing economic crops other than coconuts. The soils had been degraded by fires and oceanic storms, and their composition was up to 90 percent calcium carbonate. The creative Canadian soil ecologist, Stuart Hill, who was with us, believed that the island's soils could be made productive through the use of compost. Compost can be used to refurbish soils or can even function as a soil substitute as it mimics the cation exchange of good soils and is a very stable form of organic matter. Compost can bring other gifts to coral islands. It releases plant hormones, especially cytokinins, which in turn stimulate plants to produce larger and more branched roots. Compost is also a prime substrate for nitrogen-fixing bacteria and blue-green algae, thereby providing atmospheric nitrogen for plants. The blue-green algae are an excellent source of nutrients, too.

Stuart Hill was able to show us that compost can play one other crucial role in alkaline soils. It releases organic acids which, if applied at the right time in the decomposition process, neutralize the soil. As a result of his work, vegetables and fruits are now being grown to diversify the diet of islanders. Stuart found that the island lacked several essential minerals, specifically manganese, boron, and iron, which initially had to be imported. Our long-term strategy would be to search for local oceanic creatures that concentrate these substances which would be added to the compost. Nutrient independence is an important objective, particularly for regions of the world where foreign exchange is scarce or nonexistent.

The teachings from Java and the experiments on the Seychelles are but two of the examples which informed the work at New Alchemy and, since 1980, our newest organization, Ocean Arks International.

In fact, our ecological technologies all borrow their design features from a blend of ecosystem knowledge, materials science, and the wisdom of the Javanese farmer or the skills of the ancient Mayans of Central America, who with their chinampa or "floating" agriculture fed densely settled cities. One such technology is the aquatic farming module. The development of this technology began under my direction at the New Alchemy Institute in 1974. In summary, an aquatic farming module is a translucent solar energy absorbing cylinder up to 1,000 gallon (3,785 liter) capacity, which is filled with water and seeded with over a dozen species of algae and a complement of microscopic organisms. Within these cylinders phytoplankton feeding and omnivorous fishes are cultured at very high densities. The species selected depend upon climate, region, and market opportunities: the range of species which we have studied is broad, including African tilapia, Chinese carps and North American catfish, and trout.

Dense populations, up to 1 fish per 2 gallons (1 fish per 7.6 liters) of actively growing fish, produce high levels of waste nutrients beyond the capability of the ecosystem to take them up. The module eliminates these nutrients in four ways: i) fish growth; ii) plankton proliferation; iii) partially digested algae which flocculate out and settle to the bottom which can then be periodically discharged through a valve to fertilize and irrigate the surrounding horticulture; and iv) a modern "chinampa" system, the uptake of nutrients by vegetable crops rafted on the cylinder surface. In this case the root systems of the plants take up the nutrients before they reach toxic levels and secondarily capture detritus and function as living filters purifying the water.

These modules can be productive with fish yields, depending on species and supplemental feeding rates, of over 250 pounds of fish (113.5 K) annually in a 25 square foot (2.3 square meter) area. At the same time each unit can produce eighteen heads of lettuce weekly for an annual production of over 900 lettuce heads. Tomatoes and cucumber crops can also be cultured on the surface for even higher economic yields. The modules have the additional beneficial attributes of being water conserving. Evaporation is almost eliminated from the surface, so that makeup water rates are based upon plant evapotranspiration and the amount of module water released to irrigate and fertilize the adjacent area.

Aquatic farming modules are an agro-ecology that require initial seed capital to construct and establish, but to a large extent they are a substitute for heavy tillage, harvest, fertilizer, and irrigation equipment that would otherwise have to be used to establish and operate a farm. Not only are the modules space conserving and less costly, they can be employed in urban centers, within greenhouses in northern climates, and as a key ingredient in the process of restoring damaged environments.

Within a given land restoration project the modules could be established in rows in the most highly degraded areas. Young trees on the shaded side of the cylinders would be planted and subsequently nurtured by the periodic release of water and nutrients. On the sunny side of the modules a variety of short-term economic crops should be established to add to the produce from the module. The module-based agriculture would provide the skilled labor pool to tend the emerging ecosystems.

The aquatic farming module approach, when used for ecosystem reclamation, need not be static in the sense that the modules, having fed and watered the newly emergent vegetation including trees through their most vulnerable stages, could be shifted to new locations to repeat the process. In this way the short-cycle biotechnology could spread its benefits to surrounding ecosystems over a larger geographic area.

A related ecological technology has been developed for arid regions. For arid environments, including the Atlantic coast of Morocco, we have developed a bioshelter system to assist with ecological diversification. The bioshelter is a transparent climatic envelope or greenhouse structure which houses the fish and vegetable modules. Our prototype is a circular geodesic structure. It functions as a solar still and as an "embryo" for the early stages of the ecological diversification process. These bioshelters can operate even where there is no fresh water. In this extreme case the aquaculture modules are placed inside the climatic envelope and water from the sea is pumped through them. During the day the structure heats up and the temperature differential between the sea water in the tanks and the air is great enough to cause the tanks to sweat fresh water, irrigating the ground around them. Tree seedlings are then planted into this moist zone. At night the moisture-laden air cools to the desert sky. Water droplets form over the interior skin of the climatic envelope. With the prototype structure we found that drumming on the structure's membrane in the early morning caused it to "rain" inside. This allowed the whole interior to be planted. Also the process permits drought tolerant trees to be established around the outer periphery of the structure. Inside marine fish and crustacea such as mullet and shrimp can be cultured to form the basis of an economy. After a few years the original cluster of climatic envelopes can be moved to a new location to repeat the cycle, leaving an established semiarid agro-ecosystem behind.

These are two biotechnological examples drawn from a range of options that could help reverse environmental degradation and restore diversity and bounty to a region. These advanced technologies may well prove to be essential tools in creating sustainable environments.

Most modern societies are faced with the crisis of waste accumulation. The natural world is threatened by our inability to integrate our agriculture and industry within the great planetary cycles. Industrial cultures are cancerous, yet need not be. For me, the cleansing of water is one point of intervention. Sewage treatment plants, as an example, are expensive but do not purify water. They kill the "bugs" and remove the solids, when they work, which is not often enough. They do not remove nutrients or toxic materials. This does not have to be so if wastes are seen as resources out of place and concepts like total resource recovery underly the design of waste purification systems.

Ecologically based resource recovery can alter the economics of recycling. Sewage treatment plants are a financial drain on communities, whereas ecosystem water purification could be the basis for economic activity. Sewage can be made into drinking water and the byproducts of the process can have economic value. To demonstrate this I have been involved in the design and development of a solar aquatic waste treatment facility. It is part of a joint venture among a governmental body, the Narragansett Bay Commission; our research organization, Ocean Arks International; and the Four Elements Corporation, a new company established to take advanced ecological concepts into the marketplace.

The Narragansett Bay Commission of Rhode Island operates the city of Providence's huge sewage treatment facility. It is also concerned with protecting Narragansett Bay and its abundant but highly threatened marine resources. Their sewage plant does not remove nutrients or toxins other than those in the sediments. Our solar aquatic waste treatment facility (SAWT) is designed to remove these on a demonstration scale of 50,000 gallons a day. It is also designed to produce marketable by-products ranging from flowers to fish. It will serve double duty as a hatchery for over half a million striped bass a year. Striped bass is a marine fish whose populations have collapsed because spawning grounds and nurseries have been destroyed by pollution.

The SAWT facility is comprised of a solar-heated greenhouse within which are two parallel streams separated by 180 of the aquatic farming modules previously described. The modules trap and store solar energy creating a year-round semitropical environment inside the building. They also serve to house and nurture the young striped bass. The sewage enters at one end of the structure, and over a five-day period flows slowly through the facility. The two streams contain four sequentially arranged aquatic ecosystems, each with an essential task in the purification process. They all house biologically active food chains fed initially by the sewage.

An overview of the process is as follows: The sewage is pretreated with ultraviolet sterilization, then charged with oxygen through aeration. Introduced air is essential at each stage. The first ecosystem has an algae base, algae being the penultimate utilizers of nitrogen, phosphorous, and other nutrients. The second ecosystem is dominated by floating aquatic plants, including water hyacinths, which trap the upstream algae in their filamentous roots. They also continue to remove nutrients and take up toxic materials. The city of San Diego has found that water hyacinths remove most organic solvents as well as heavy metals. San Diego, along with the space agency NASA, has pioneered water hyacinth purification of sewage. The third ecosystem is made up of clear water with artificial habitats on the bottom in which microscopic shrimplike animals graze on the algae and bacteria resident on the attached substrate. They are fed upon by mosquito fish and fundulus which in their turn are fed to the bass in the adjacent aquatic farming modules. The fourth and final ecosystem is a marsh comprised of reeds and bulrushes planted in a gravel filter. These higher plants, up to twenty to thirty feet tall in the greenhouse, remove any remaining organisms and toxins. They also polish the water. Reed and bulrush waste treatment was the brain child of Dr. Kaethe Seidel of the Max Planck Institute in Germany. She discovered that marsh plants could tansform sewage into potable quality water. Her research findings have given new meaning to the protection of wild marshes. After the water passes through the marsh filter in the SAWT facility, it is ready for reuse. In the case of the prototype it will be used for local industrial needs.

The solar aquatic waste treatment demonstrates the value of ecological integration and illustrates how nature's bounty can be applied to human needs. Most of the third world is plagued with sewage-born diseases and cannot afford industrial waste treatment. Even if they could, they would be robbed of precious resources. SAWTs can be designed to control disease and also to serve as epicenters for the production of fertilizers and the cultivation of plant materials including trees for reforestation.

We expect to build the prototype in 1987. Our research leads us to be confident it will work in the cold New England climate. It is our hope that it will be the seed for a new commitment to caring for water as our most precious and elemental resource. Stewardship needs to be extended to our ground waters, lakes and streams, and the ocean. It is our sacred trust to the planet, to Gaia, to shift our values so that our first order of business is to cleanse the waters, protect the soils, and tend to the trees.

I have sketched only a few of the ideas and technologies which are derived from ecology. I am aware that ours is a world of violence, hunger, environmental degradation, and inequities. For most of us points of action and intervention on behalf of the planet and ourselves may be hard to find. But I believe this will change if our economies become ecological. Work and stewardship will be one. An ecological economic order has the intrinsic potential to allow each culture to explore the new frontier in its own way so that some of the old divisions between peoples and places can be reduced. Fritz Schumacher worked for greater equity and justice, and so must we.

Notes

This paper was originally presented as the annual lecture of the Schumacher Society, 1985, and is printed with permission of the Society. Information on the work of the Society may be obtained by writing them at RD 3, Box 76A, Great Barrington, MA 01230.

FURTHER REFERENCES

1. Nancy Jack Todd and John Todd, *Bioshelters, Ocean Arks, City Farming: Ecology as the Basis of Design* (San Francisco, Sierra Club Books, 1984).
2. John Todd, "Planetary Healing," *Annals of Earth Stewardship,* 1983, Vol. 1, No. 1, pp. 7-9.
3. John Todd, "The Practice of Stewardship," in *Meeting the Expectations of the Land,* Wes Jackson, Wendell Berry, and Bruce Coleman, eds. (San Francisco, North Point Press, 1984), Chapter 12.
4. Ron Zweig, "An Integrated Fish Culture Hydroponic Vegetable Production System," *Aquaculture Magazine,* 1986, Vol. 12, No. 3, pp. 34-40.

8

HAZEL HENDERSON

A Guide to Riding the Tiger of Change

The Three Zones of Transition

IT IS HARDLY news to anyone that industrial societies are undergoing massive structural changes and realigning themselves in a process of economic and technological globalization. Today, this planetization process is visibly accelerating and three distinct zones of this unprecedented transition can be mapped to help decision-makers negotiate this unfamiliar terrain: 1. The Breakdown Zone, 2. The Fibrillation Zone, and 3. The Breakthrough Zone.

Since all of us live in one or more of these zones and few forecasting methods are broad enough to capture such overall dynamics, we must shift our attention from modeling *content*, i.e., the daily quantification of events, data, to modeling the wider *context* of these events and the overall *processes* involved. Attempting this heroic modeling task makes amateurs of us all, and yet it is crucial in creating the new conceptual tools required if we are to learn to interpret these events and to ride the tiger of change.

With this overall context of accelerating globalization, evident in areas from banking and finance, satellite telecommunications, computerization, air transportation, militarization, and the speedup of technological innovation, we can also expect increasing turbulence and new instabilities. Further, we should expect that more of the changes we see are *irreversible,* while taking note that most of our conceptual tools for mapping them—such as economics and conventional scientific approaches—are still based on Newton's ideas of mechanics and reversible models of locomotion in a clockwork universe. Therefore, we can also expect accelerating "future shock" (to use Alvin Toffler's term), even in formerly stable areas of our personal and political lives and institutions. All this will occur in the context of swifter and larger shifts in environmental conditions, as new thresholds are overridden, as for example where increased carbon dioxide in the atmosphere is now producing more climatic variability. Another effect to observe will be the *ambivalence* of these events, with more confusion and conflicting interpretations by scientists, governments, and media, i.e., the "is it good news or bad news" syndrome. Our three-zone map may help give us a pegboard to sort things out for ourselves and pinpoint where we are in the picture. Since all three zones coexist simultaneously, we might also remember that word maps, such as this article, are less effective than pictorial maps and that even then, flat surface maps are less representational than a three-dimensional globe—the real stage on which the three transitions are occurring.

Zone 1 The Breakdown Zone

In Zone 1 many of us feel that our lives and jobs are stultified, or that we are stuck in an unresponsive bureaucratic institution or corporation. This is natural in a time of change, since individuals always learn faster than institutions. In fact, institutions often rigidify, resisting change until they become brittle and shatter, while others simply stagnate or decay. Thus, this Breakdown Zone is where society and its obsolescent institutions are de-structuring. We need not panic, since de-structuring is a natural process like composting, creating a rich new soil for regeneration. In fact, Nature shows us how some species actually regress to an earlier, larval stage in their development when their adult form has become too rigid and ill-adapted. This process, paedomorphosis, allows the younger, less-structured (and therefore more adaptable) form to carry on the species. So it may help us to see Zone 1 as containing these "seeds," and remember that paedomorphosis leads to the many metamorphoses we will find in Zone 3, the Breakthrough Zone.

In Zone 1 it is not only institutional forms, cities, suburbs, and rural areas that are de-structuring, but also cultural and political forms and value systems underlying them. For example, our culture and those of most other industrial societies are in a state of confusion as they shift to the not-yet-defined "post-industrial" phase. The Soviets and other socialist societies experiment with marketplace heresies to overcome lack of incentive and enforced cooperation, while in the U.S.A. we yearn for less individualistic, dog-eat-dog competition and retreat into our churches, new religions, and cults in search of community and kindness. Both capitalism and communism are revealed as superficial ideologies concerned merely with methods of production and distribution, rather than deeply sustaining philosophies of life. Similarly, imposing one or the other of these two outdated European styles of industrialism on the rest of the world is failing, from Africa and Asia to South and Central America. China seems to be finding a "third way," or as Deng Xiaoping is quoted, "When there are mice in the house, a black cat is as good as a white cat." Enforcing industrialism as a single model for development is now inappropriate for the world's rich variety of diverse societies, each with its own unique expression to offer the global melting pot.

Thus, Zone 1 is also a war zone, as conflicting cultures, ideologies, and religions clash in the new global village—adding to the enduring competition between nations over territory and resources. Even if accidental nuclear exchanges are avoided, we can expect proliferation of the proxy wars, such as those in Central America and Africa, and other Third World countries. Such overt and covert violence, together with continuing inequities and injustice will continue to fuel the revolts, insurrections, and terrorism, while guerilla strategies and suitcase bombs will continue to be the natural response to military might and Star Wars.

Zone 1 is also the "accident zone" and the zone of "slow motion crises," such as pollution. Accidents, from Three Mile Island, Times Beach, Love Canal, Bhopal, the Challenger explosion, Chernobyl, and the Rhine

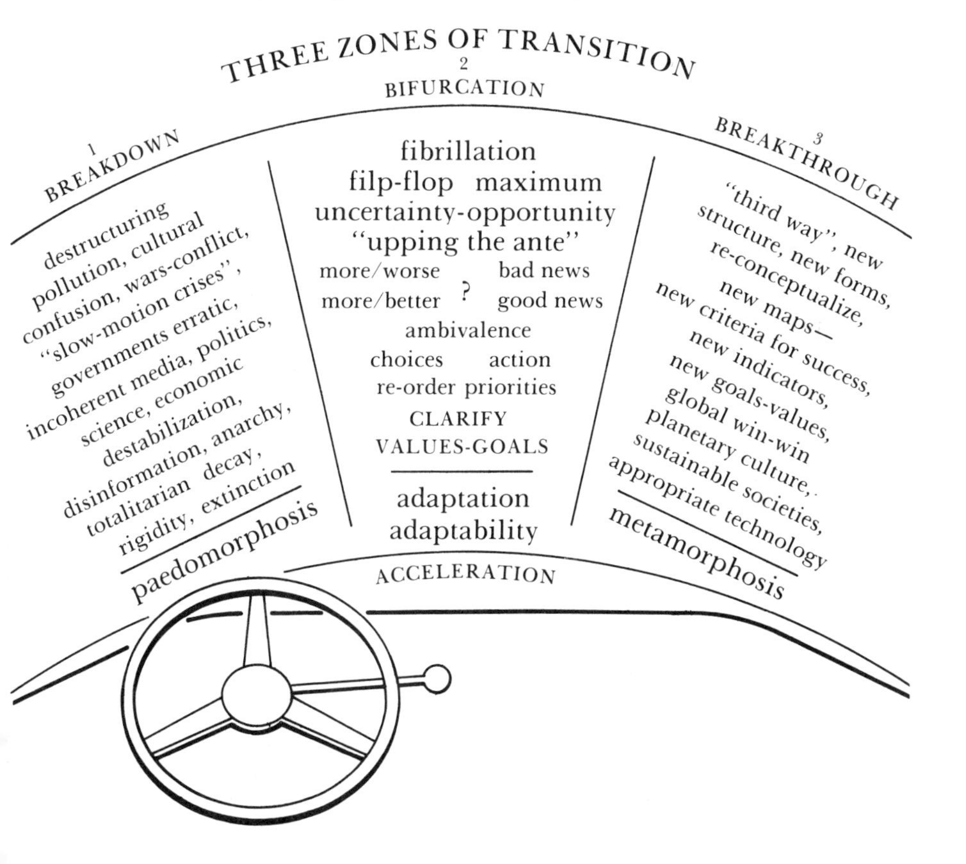

spills are all effects to be expected as we humans attempt to manage and coordinate ever larger and more complex organizations and powerful technologies. Slow motion crises to watch include increasing loss of forests due to acid rain, spreading deserts in the Sahel and in over-irrigated valleys in California, climatic warming and variability due to the "greenhouse effect" of rising carbon dioxide leading to rises in the sea level, as well as the lowering of the U.S. water table and irreversible pollution of ground water with toxic wastes.

The political arena of Zone 1 is best summed up as "the politics of the last hurrah," i.e., maladaptation to change, where governments of all ideological stripes rigidify and try to defend their borders against the waves of globalization now swamping their cherished national "sovereignty." This is most evident in the economic sphere, as somewhere between $150 and $500 billion (no one is sure) of footloose money sloshes around the planet every 24 hours and electronically transferred funds are deployed by the new breed of 24-hour asset managers playing such new games as program trading in today's "global casino" so well described in the pages of *Business Week*. Information has become money and money has become information as I described in *The Politics of the Solar Age* (1981). As the global "fast lane" speeds up, money loses its meaning and ceases to function as a viable means of keeping track or score of the game. It is in this light that Peter Drucker contends in *Foreign Affairs* (Spring 1986) that the commodity economy has "uncoupled" from the industrial economy, the industrial economy has uncoupled from the employment economy, and world trade has uncoupled from world financial flows. In his otherwise insightful article, Drucker does not go far enough by singling out these areas. In staying within the traditional paradigm of economic and money-based analysis, he misses the non-money denominated sectors of total productivity and fails to see the extent to which this new global funny

TOTAL PRODUCTIVE SYSTEM OF AN INDUSTRIAL SOCIETY
(THREE LAYER CAKE WITH ICING)

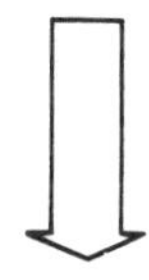

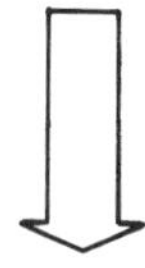

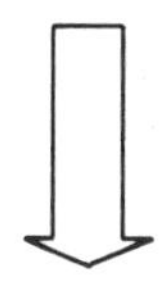

money game (or symbol system as Drucker terms it) now has very little to do with the realities of *any* sector of real-world production, consumption, investment, or trade, *nor of any real geographic region or ecosystem on the planet.*

Meanwhile, politicians wrestle with domestic unemployment, trade, retraining, and industrial policies (a hopelessly outdated concept), all of which deal with real geography and real people, and implementing such policies must take *years* of preparation and building. Yet all such domestic plans, however well laid and executed, are destabilized *daily,* as the currency exchange markets open each morning in London, New York, and Tokyo. Treaties and economic theories alike, addressed to *inter*national competition and trade policies, or to domestic unemployment, inflation, deficits, or interest rates, are all swept along by this rising tide of financial flows, as well as Third World debt, bouncing currencies and oil prices—all indicators of the need for *global* economic cooperation and a new Bretton Woods to write the necessary "win-win" rules for operating the new global eco-

WORLD TRADE ROLLER COASTER
SCHEMATIC

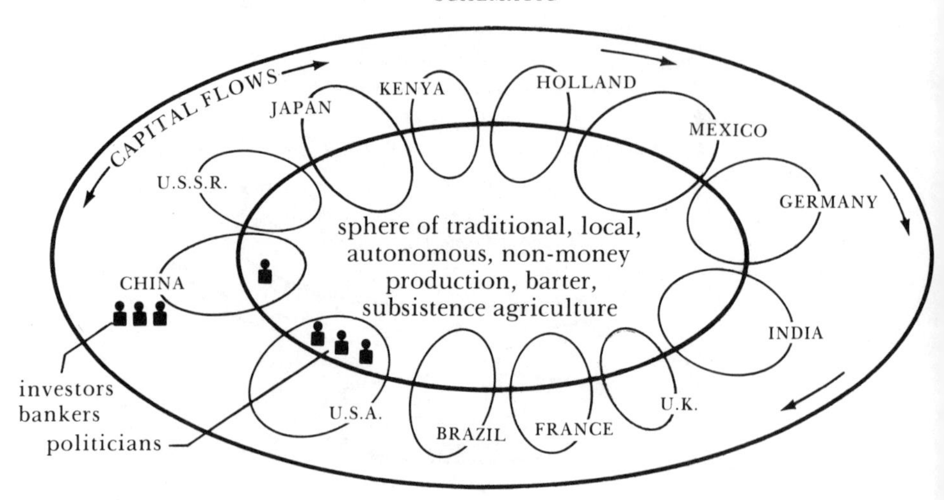

nomic "commons," i.e., as global "common property" of all the players. When any market expands toward globalization, it inevitably becomes a "commons" (a term derived from England's feudal village greens, or "commons," where every villager could graze his herds). In markets, competitive, zero-sum "win-lose" games prevail, while in "commons" unless cooperative "win-win" rules are substituted, then all players lose and the "commons" is destroyed for everyone. (See *Science,* Dec. 13, 1968, p. 1243.)

Some governments respond to Zone 1 conditions by reconceptualizing this new "global commons," while others either rigidify, try to turn the clock back, attempt diversionary military adventures, fudge the figures, or even indulge in disinformation, often confusing their own citizens by obfuscating the issues. The least adaptive political behaviors, of course, are totalitarianism or anarchy. Therefore, if you find yourself in Zone 1 too much of the time, you may recognize that it is time to assess your options, recycle your skills, and scan for opportunities to redeploy yourself and prepare for a

SPEED-UP OF WORLD TRADE ROLLER COASTER
SCHEMATIC

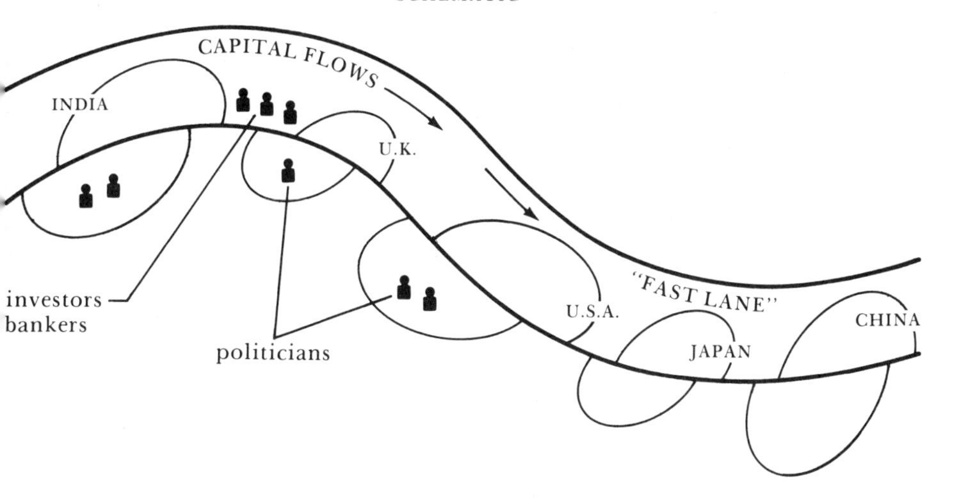

well-informed leap into Zone 3, the Breakthrough Zone. However, to accomplish this you will need to explore and negotiate Zone 2.

Zone 2 The Fibrillation Zone

Zone 2 is characterized by the term "fibrillation," as when the human heart muscle temporarily vacillates under stress, either leading to a heart attack and possible death, or shifting to another regular rhythm. Thus Zone 2 is expanding rapidly as globalization accelerates, and its atmosphere is one of "upping the ante" and a pervasive increase in risk and uncertainty. Zone 2 is a "critical mass" zone of *bifurcation* (a term used by mathematicians and those in the physical and life sciences) meaning the many modes in which a system can, or is about to change in its entirety, or state. These dynamic, organic models of changing systems include the models of "catastrophe" of French mathematician René Thom, who described seven different bifurcation modes of transformation; the "order through fluctuation" models of Nobelist chemist Ilya Prigogine of Belgium, and the "change through attraction" models of U.S. mathematician Ralph Abraham, whose computer simulations of system change processes exhibit three organic "attractors" (point attractors, periodic attractors, and chaotic attractors), which "pull" systems into new states, as do magnets. They appear very unpredictable because the minute changes they cause can give rise to very different or large results. These advances in mathematical modeling are best summed up for lay readers by Marilyn Ferguson in her scientific newsletter, *Brain-Mind Bulletin* (Los Angeles). From these models, it is possible to see how the destructuring processes of Zone 1 give rise to the uncertainties and the maximum number of opportunities to shift gears, reconceptualize, redesign and restructure, i.e., to ride the tiger of change into the third, Breakthrough Zone.

Zone 2 is also characterized by "flip-flop" processes, as whole systems enter this bifurcation zone of transformation, when they are poised on the "cusp" of these shifts in state. For example, a corporation in a state of rapid growth is suddenly confronted with a key choice, which when made, will either propel it into bankruptcy or to new markets in a restructured form. High-risk strategies are often most effective, while doing nothing can be the most dangerous "action." In Zone 2 more individuals, institutions, and nations must make choices because they are nearing thresholds and pushing against their margins and boundary conditions. For example, today's nation-states' boundaries and nationalistic belief systems have become dysfunctional. Giving up some "national sovereignty," e.g., over their domestic economies, is risky, but less risky than going it alone. Thus, in Zone 2, choices and actions are required—but unless the situation is also reconceptualized and remapped, the action may be maladaptive and relegate the system or person back to Zone 1. Thus, Zone 2 demands the most clear-eyed and rigorous reexamination of assumptions, priorities, goals, and the very values that underlie them, since values are the basic driving force in all technical, economic, and political systems. This reexamination is in itself a high-risk task, because old verities and old institutions must be challenged, which initially contributes to the destructuring process of Zone 1. Yet the price of *not* challenging the old forms is loss of leadership, those "attractors" which "pull" the system (in political terms, those with more attractive visions of the future) into its new state. For example, the Democratic Party's current disarray stems from its attempt to *emulate* Republican policies, rather than envision a new direction for the country. Republican "politics of the last hurrah" successfully repackaged Keynesianism as "supply-side" economics and flipped the "malaise" of the Carter Administration into the soaring deficits and fervid over-optimism of the "high frontier" and Star Wars. Neither party reexamined basic values, but simply sloganized our traditional values of optimism, entrepreneurship, sharing, and cooperation, without reconceptualizing our geopolitical and economic dogmas of competition, and our outdated worldview of the U.S.A. as Number One, Fortress America, dominating our rivals with military and economic superiority.

Another key aspect of Zone 2 is that we should expect more and better good news and more and worse bad news. This effect is evident in falling oil prices and is simply another indicator of systems reaching margins and thresholds of maximum stress conditions. It helps explain why incremental changes are often ineffective. Ony policies addressing *basic* causes which underlie problems can hope to succeed, while policies addressed only to ameliorating or suppressing symptoms can lead to worse results. For example, trying to deal with unstable domestic economies using narrowly focused and superficial remedies of traditional "flat earth" macroeconomic policies: inflating or deflating, regulating or deregulating, privatizing or nationalizing, raising or lowering interest rates, as if the society were a hydraulic system, can leave the patient worse, possibly irreversibly. As mentioned, since globalization has changed the game, only global agreements can work, as well as fundamental reevaluation of all economic sectors, reanalysis of data and developing of new indicators of performance beyond the simplistic model of the Gross National Product, such as Japan's Net National Welfare (NNW) or the Physical Quality of Life Index (PQLI), as I have detailed elsewhere.

Thus, Zone 2's proliferating good news and bad news becomes also evermore ambivalent, and it becomes increasingly unrealistic to view any news in such categorical, either/or terms. Media analysts can no longer help by merely reporting the events, since *interpretation is everything* and examining the underlying causes and assumptions of the actors and the audience becomes the key to deciphering the unfolding plot. For example, many futurists, including myself, have focused on the so-called "information age's" good news, and there is plenty of it: trends to greater participation, more informed citizens, decentralization, as well as the "high touch" which John Naisbitt, author of *Megatrends*, sees as balancing the less pleasant side of the high tech revolution, as well as the now widely touted labor "shortage." However, the full dimensions of the "information age" lead to more ambivalence: more efficient computerization of the military may trigger an accidental nuclear exchange; the rapid disruption of work due to automation; health and privacy effects of the computer revolution and a growing glut of raw unevaluated data; a U.S. economy characterized by "hollow corporations" which merely label and sell ever more foreign-made cars and consumer electronics, and a "services economy" of clerks and hamburger flippers whose apparent "labor shortage" to fill these low-paid tasks helps mask the millions of semi-illiterate, structurally unemployable minorities and youth—all propelled by the globalization processes of the electronically speeded global casino.

So, if you, like most of us, find yourself living most of your life in Zone 2, the best course is to dig deeper for the answers, to turn the issue or event (whether reported as "good" or "bad" news) around and look at all its facets, while surveying the widest range of interpretations offered by politicians, business leaders, unionists, academic forecasters, or futurists, summarized in such publications as *Future Survey*, of Washington, D.C. Zone 2 is the arena of trade-offs between *adaptation* and *adaptability*. If we or our institutions have become too well adapted to now vanishing conditions, we will have less in our storehouse of adaptability to meet the new conditions—the "nothing fails like success" syndrome. Anthropologists call this The Law of the Retarding Lead, and we see it operating today, where countries which are less industrialized, such as China, India, and Sri Lanka, may be able to forge ahead by taking the best from all the earlier experimenters in Europe, North America, and Japan, and leapfrogging to a Third Way, thus entering Zone 3.

Zone 3 The Breakthrough Zone

This zone of breakthroughs was almost invisible during the 1960s and 1970s, because they could not emerge until sufficient destructuring had occurred. As the Breakdown Zone widened and led to the expanding Fibrillation Zone, so too, the breakthroughs grew and became more visible: new compacts among nations, such as those bordering the Mediterranean Sea to clean up their common pollution; the treaties to guard against the militarization of outer space and to protect the South Pole from exploitation; as well as the many United Nations-sponsored conferences on Law of the Sea, and the global issues crosscutting national boundaries: food, population, health, education, habitat, renewable energy sources, and science and technology for development. These brave beginnings to create new global social management technologies were coupled with new human capabilities in unlocking the basic code of life: the DNA molecule and such advances as the eradication of smallpox. A growing awareness of our human powers and responsibilities for more appropriate uses of our scientists and technologies to extend human lifespans and potentialities, and to end hunger and disease, led to the new dialogues between rich and poor nations of the Northern and Southern Hemispheres over a more just global economy.

New sensitivites emerged to appreciate the diversity and richness of ethnic cultures and, at last, a concrete vision of planetary identity flashed from space to a whole generation of the human family.

Zone 3 is where old "problems" and "crises" are revealed as new opportunities, and the good news in the bad news becomes apparent. Even the nuclear bomb had, indeed, kept the peace for forty years, and so the new rounds of proliferation forced millions of citizens to demand arms reduction treaties and a shift of resources away from the dangerously growing militarization, toward finally dealing with poverty, disease, hunger, and war—the four horsemen of the *real* apocalypse. All through the 1970s and 1980s, citizen movements grew in all countries for peace, human rights, government and corporate accountability, and ecological sanity. Socially responsible investments and successful mutual funds, such as the Calvert Social Investment Fund of Washington, D.C. proliferated with these movements. Tyrannical regimes fell in Iran, Haiti, and the Philippines, while others tottered, including South Africa, now reeling from citizen-spurred disinvestment and the courage of its own black citizens. Similarly, the outdated regimentation of the old-style industrialism, based on inadequate understanding of human needs and potentials and with its limited awareness of Nature's crucial role in production, is now leading to more humane, participatory organizations, cooperatives, worker-owned and managed businesses, and the burgeoning of smaller businesses and entrepreneurship, as well as production methods, recycling, and recovery systems that work with Nature and within ecological tolerances. I termed this shift toward less resource-intensive, sustainable forms of production, consumption and investment, a shift toward a dawning Solar Age, an Age of Light, where humans remembered that all processes on Planet Earth are dependent on the daily flow of photons, the life-giving light from the Sun, our mother Star.

Today, we are already moving beyond the "information age," based on electronic technologies, to the Age of Light and its lightwave technologies: from lasers, fiber optics, optical scanners, and computing, to photovoltaics and many other thermal and chemical energy-conversion processes based on a deeper understanding of Nature and modeling her processes—from solar collectors based on the chloroplasts in every green leaf to biotechnologies based on the genetic code, still in their moral infancy as they explore these new Faustian powers. As United Nations Assistant Secretary General Robert Muller reminds us in his *New Genesis* (1984), we are a very young species, in terms of our planet's development, and we have in our very brief history learned a great deal, and as long as we refuse to panic or despair, we may yet learn the lessons of globalization now upon us. As I have elaborated elsewhere, our planet is a perfectly designed programmed learning environment, akin to one of psychologist B. F. Skinner's famous boxes—providing us with all the lessons and both positive and negative feedbacks needed to nudge us along the path.

We see the learning now occurring through the "crises" of our costly, mechanized, chemical, and energy-dependent agriculture and its massive production of monocultured crops now glutting world markets. As agriculture restructures under the stress of globalization the same diversification and entrepreneurship now restructuring industrial sectors is at work. The future lies in lower-cost, lower-input forms of agriculture, in smaller-scale, and "boutique" farms, in new crops, from jojoba and guayule to ethnic fruits and vegetables, specialty and organically grown foods, fish farms, and genetically engineered varieties, tolerant of poor soils, excessive salt, and lack of water.

Here too, we see that this planetary storehouse of genetic diversity is a "commons," just as are the oceans and the air we breathe. Therefore, we must also conclude global compacts as rapidly as possible to move toward "win-win" rules to manage these precious resources cooperatively, for the benefit of all the human family, rather than in the obsolete self-destructive, competitive mode of today's biotechnology sector, whose research base has been underwritten by public funding and taxes. Such technologies are too precious and potentially hazardous to be left to the mercies of a few unscrupulous or careless companies, which can put others in jeopardy and foreclose the options of future generations. Similarly, doctrinaire, laissez-faire assumptions are now hampering the wider development of the computer sector, where competitive, zero-sum rules are creating a tower of babel of incompatibilities, preventing wider use of computers in the global, networking modes to which they are naturally suited, based on the random-access model of telephone systems, as I elaborated in "Computers: Hardware of Democracy" (*Forum*, Fall 1969, and *Harvard Business Review*, May-June 1971). Another global information "commons" has emerged. More than thirty companies have already joined a consortium, the Corporation for Open Systems, which is trying to promulgate the new "win-win" rules: a set of worldwide, common standards, while France has led the way by offering free home terminals as a "common good" so that all householders can benefit from access, thus opening up a vast market for services.

Even the threat of global economic chaos is forcing governments and economic advisors to throw out old ideologies and address the new agenda of stabilizing currencies and financial flows, and seek more realistic *niches* of true comparative advantage and symbiosis. We see that head-on competition to produce a narrow range of goods in already saturating markets is now a destructive exercise in putting lower wages and further ecological destruction on the marketplace's auction block, and turning economic activities into a global, zero-sum behavioral sink.

Thus, Zone 3 involves not only breakthroughs, restructuring, new forms and adaptations, but also a broad "politics of reconceptualization" of all the basic assumptions and conditions underlying the "problems" and "crises" of Zone 2. Knowledge is restructured from old, single disciplines, such as economics, into new, trans-disciplinary policy tools, for example, from ma-

DIFFERING PERCEPTIONS, ASSUMPTIONS, AND FORECASTING STYLES BETWEEN ECONOMISTS AND FUTURISTS

ECONOMISTS	FUTURISTS
Forecast from past data, extrapolating trends	Construct "what if?" scenarios Trends are not destiny
Now also use optimistic, pessimistic forecasts	Identify "preferred futures" Plot trends for cross-impacts
Change seen as *dis*-equilibrium (i.e., equilibrium assumed: all other things equal)	Fundamental change assumed (transformation assumed)
"Normal" conditions will return	No such thing as "normal" conditions in complex systems
Reactive (invisible hand assumed to control)	Proactive (focus on human choices and responsibilities)
Linear reasoning Reversible models	Non-linear reasoning Irreversible models, evolutionary
Inorganic system models	Living system, organic models
Focus on "hard" sciences & data	Focus on life sciences, social sciences "soft," fuzzy data, indeterminacy
Deterministic, reductionist, Analytical	Holistic, synthesis, seeks synergy
Short-term focus (e.g., discount rates in cost/benefit analysis)	Long-term focus, intergenerational costs, benefits, and trade offs
Data on non-economic, non-monetarized sectors seen as "externalities" (e.g., voluntary, community sectors, unpaid production, environmental resources)	Includes data on social, voluntary unpaid productivity, changing values lifestyles, environmental conditions map contexts, external variables. (Use post-economic models: technology assessment, environmental impact, social impact, studies)

DIFFERING PERCEPTIONS, ASSUMPTIONS, AND FORECASTING STYLES BETWEEN ECONOMISTS AND FUTURISTS

ECONOMISTS	FUTURISTS
Methods tend to amplify existing trends (e.g., Wall Street Psychology: "herd instinct" in investing, technologies, economic development)	*Methods* "contrarian," (e.g., look for anomalies, check biases in perceptions, cultural norms) Identify potentialities that are latent
Entrepreneurial when "market" is identified	Socially entrepreneurial (Schwartz) (e.g., envision future needs, create new markets)
Precise, quantitative forecasts (e.g., Gross Nat. Product for next quarter of year annual focus)	Qualitative focus (e.g., year 2000 studies, anticipatory democracy) Data from multiple sources, plot interacting variables, trends in long-term global contexts

croeconomics to "post-economic" policy studies including technology assessment, environmental impact statements, futures studies, scenario-building, cross-impact analyses, risk assessments, social impact studies, and systems research—all with global, rather than national frameworks. This composting and recycling of our knowledge is already leading to new maps of such unnoticed territory as the informal, non-money denominated sectors of production, services, and investment, which match and often subsidize the economists' more familiar money-denominated, GNP-measured half of the picture (as I have outlined elsewhere, see page 149).

As these new maps clarify the new terrain, new criteria for "success" and new indicators and measures of performance and "development" are emerging, for example, the GNP is giving way slowly, in many government agencies and academic textbooks to broader indicators, such as Japan's Net National Welfare (NNW), and the Overseas Development Council of Washington, D.C.'s P.Q.L.I., mentioned earlier, and the Basic Human Needs (BHN) indicator, developed by the United Nations' Environment Program. Using such indicators, a very different picture emerges, and such countries as Sri Lanka and China are highlighted for achieving progress in health, education, shelter, and environment, as well as mere growth of per-capita-averaged money income (which often masks severe inequities in distribution). As these indicators take hold, it becomes clear that countries such as China are achieving their successes partly due to the shifting priorities, and China's military expenditures have been reduced each of the past thirteen years. Similarly, Japan's success is due in part to her concentration on serving civilian markets, rather than joining the deadly, costly competition of the arms race.

In Zone 3, we also see that the old "either/or" debate gives way to a "yin/yang" view of complimentarity. For example, the debate moves beyond the *either* competition *or* cooperation argument, to the understanding that both these equally important principles are operating simultaneously and at every level in all human societies and in Nature. In many countries, we see also that both politics and economics are moving beyond the one dimensional, left-right perspective to a more fully dimensioned debate of the major factors that now must be included: 1. globalization; 2. ecology; 3. the non-money denominated sectors of production, exchange, and investment (for example, 25% of all world trade is now conducted in barter) as well as the cooperative, voluntary sectors and our changing lifestyles and values; and 4. the future long-term costs, benefits, and trade-offs and intergenerational risks and rewards of our short-term actions.

Already Zone 3 is replete with new concepts: such ideas as a *Pacific Shift* of economic and cultural leadership, as expounded by historian William Irwin Thompson in his 1986 book, and the ubiquitous images of emerging planetary paradigms and cultural identity; the concepts of sustainable forms of production, renewable resources, appropriate technologies, and the new economics (or "ecologics," as some term it) of the carrying capacities of various ecosystems, not to mention the exciting new views of human nature and potentials growing out of brain and mind research, summed up in Dr. Jean Houston's *The Possible Human*. Zone 3 is also replete with models and examples of "win-win" breakthrough strategies. However, they appear as "insignificant" to Newtonian-trained scientists with single-disciplined or "clockwork" models, rather than the "attractors" of the new, organic systems models, such as those mentioned earlier. Similarly, most statistical "cameras" are still focused for vanishing phenomena of a more discrete, static, orderly world of the past. For example, social policy in the U.S. is still largely based on the old model of the single-breadwinner, nuclear family, with stay-at-home wife and two children, even though such families now account for only about 10 percent of the total. Similarly, economic statistics ignore the flows of services in world trade, now enormous, while economic models cannot embrace the ubiquitous new commodity: information, which is not scarce, and therefore comforms to "win-win" rules, rather than zero-sum competition, as I have described elsewhere.

Today, it is well for our mental health to remember that the supercharged atmosphere we are experiencing as we move further into the flip-flop modes of Zone 2, is still the focus of most academics, statisticians, and mass media. Thus the breakthroughs are continually overlooked or drowned out by the saturation reporting of the daily shocks, threats, confrontations, and senseless violence of Zone 1, while the opportunities and choices of Zone 2 are under-reported or misinterpreted. For example, the widely covered insurance "crisis" now being blamed on either insurance companies or lawyers and the jury system, is a golden opportunity to examine the limits of inherently risky technologies to address tasks that can be accomplished in other less risky ways; the Newtonian, "clockwork" assumptions underlying most of our models for assessing risks and probabilities; *and* the overall social values implicit in the current insurance system (see my "Risk, Uncertainty and Economic Futures," *Best's Review*, May 1978).

The quiet building and restructuring taking place in Zone 3 is "slow motion good news" and cannot be summed up in 30-second pictures between commercial breaks on half-hour news shows, and yet, it is vastly more important to our future than most of today's "photo opportunity journalism." For example, we hear of all the giant corporations that fail or close plants, while most of the 700,000 small new companies formed each year go unnoticed because the Census Bureau does not count companies with fewer than 20 employees. As the acceleration of change increases, "news" and "facts" will fall further behind the changing scene. ABC-TV's Ted Koppel often performs diplomatic functions and ventilates national confrontations and "hot-spots" on Nightline, before the State Department or the UN can get to them. U.S. Radio's "Global Town Meeting" broadcasts have reached many millions, amplifying people-to-people exchanges by linking up dozens of cities around the world in live discussions of peace strategies: U.S.-U.S.S.R. cooperative space missions; economic conversion to civilian production, and citizens' activities. The new models of global radio "town meetings" and TV telethons for hunger are expanding rapidly, for example, in such marathons for global peace and cooperation as The First Earth Run, Hands Across America, and the UN Day of Peace, as well as the TV teleconferences linking the citizens of diverse cities in face-to-face, "space-bridge" satellite-linked meetings. The networks of the 1970s are now manifesting, and the agendas for the new planetary societies are clear: in such documents as the World Resources Institute's *Global Possible* proposals; Worldwatch Institute's *State of the World Reports*; in the UN's Declaration of Human Rights and the programs of the World Health Organization, UNICEF, and other agencies; and the China 2000, Africa 2000, and some fifteen other such studies in other countries.

Here, once again, we find that many "crises" and "problems" are opportunities. For example, we find that the much-bemoaned world population "problem" may actually be stabilized by *saving lives.* By preventing millions of needless early infant deaths from fatal diarrhea in many countries, the World Health Organization tackled *both* tasks, by the simple, swift, inexpensive remedy of administering a drinking solution of water, glucose, and salt. In so doing, this oral rehydration therapy (ORT) has *reduced* birthrates, rather than increasing them, as Newtonian-oriented studies expected. The really good news is that so many solutions are turning out to be simple and inexpensive, rather than requiring massive, costly new technologies. When "crises" and "problems" are fundamentally reexamined, solutions often arise in the rethinking process, as in the "lateral thinking" and creativity exercises used by many organizational development and transformation theorists. For example, massive, costly high-tech medical systems to "cure" disease now are giving way to less costly remedies: healthier lifestyles, less and better nutrition, more physical activity, education, and prevention, as well as new understanding of the beneficial effects of less stress and more positive outlooks on life.

As our mass media begin to understand their role as the nervous system of the new body-politic of the human family, they may also search out and interpret the events and opportunities of Zones 2 and 3, thus reducing general stress levels and panic reactions, while amplifying our knowledge of all the healthy choices open to us. Clearly, the human species is at a new evolutionary juncture, and is undergoing the timeless drama of all species: the play between adaptation and adaptability, between maladaptation, paedomorphosis, learning, transformation, and metamorphosis. As we deal with the heightened stakes of the Fibrillation Zone with all its unavoidable choices, we can all do our part in taking the millions of necessary small steps and wise decisions, most of which we know intuitively, which together will amplify the "attractors" leading us toward the further expansion of the territory of the Breakthrough Zone. The vision of successful globalization will govern the "win-win" politics of building an equitable, culturally diverse, ecologically harmonious and therefore peaceful planet. In this all-embracing context, all our individual self-interests become coterminous: in the self-interest of our now truly interdependent human family, as we emerge into The Age of Light.

Note

Most of the ideas discussed in this article are treated at length in my *Creating Alternative Futures* (1978) and *The Politics of the Solar Age: Alternatives to Economics* (1981) and in "Post-Economic Policies for Post-Industrial Societies," *Re-vision*, Winter 1984, Cambridge, Mass.

9

WILLIAM IRWIN THOMPSON

Gaia and the Politics of Life

A Program for the Nineties?

I. Of Conscious Purpose and Unconscious Polities

IN HIS ESSAY, "The Effects of Conscious Purpose on Human Adaptation," Gregory Bateson showed how the conscious purpose of a society, expressed in its economic policies, had very little knowledge of its biological life within an environment.[1] A society did not *know* what it was *doing*, or, in another way of saying it, its political interpretation of life was less than its full existence in an ecology. All that was left over when the conscious interpretation of activity was subtracted, constituted the virtual existence of the organism embedded in the environment. This unconscious transformational activity at the membrane between the organism and the environment was, for Bateson, still the expression of a kind of Mind, and Bateson's last work in his life was to try to explore just this relationship between Mind and Nature.

The nervous system, in Bateson's descriptions, only reports on its *products* and not its *processes*; similarly, society only reports on its industrial products and not on the condition of its ecological processes. Economists will describe the conscious structure of a society in the ratiocinative language of quantitative measurement, and this conscious description is called the Gross National Product. The unconscious process, the actual life of the culture within an ecology, is peripheral to the value system and is experienced only as incidental pollution. It is paradoxical that although the GNP is invisible, and pollution is most visible, the abstraction is taken for concrete reality and the sensuous experience dismissed to the margins of society, where it is picked up by such marginal elements as artists, philosophers, and the generally disaffected.

If a political policy is unsound, one discovers it through noise. Noise is an expression of the ignored and the unknown, of the irrelevant and the unvalued. As the noise builds up it reaches a point in which it overwhelms the signal, and then one gets a reversal in which the noise begins to be heard as information and the old signals fade into a background hum, a musak of buzzwords and archaic rhetoric.

Pollution itself is a form of noise in the transmission of human conscious purpose into the wild. At the beginnings of civilization, such noise is ignored, and only now are such disturbances as soil-loss, water poisoning, and atmospheric contamination becoming truly disturbing. As this noise continues to build up it will reach the point when it will overwhelm the signal and the industrial rhetoric will become a noise mechanically recited by people still invoking a historical envelopment that is no longer the actual historical environment.

At the point that noise begins to be heard as information, one begins to get a sense that noise is actually systemic rather than random and that it constitutes a form of echo or shadow to the unrecognized civilizational system. Here we are no longer talking about an intellectual unconscious, or *episteme*, (chez Foucault), but a civilizational unconscious. The industrial nation-state with its GNP is the conscious polity, but the unconscious polity with its noise and pollution is the gaseous and nebular shape of things to come. It is the cloud of Chernobyl that in its movement does not recognize national boundaries.

Our unconscious polity is a biome we experience as U.S.A.-Mexico-Canada. For the life of a biome, the boundaries of the nation-state are illusory abstractions. Our southern border is melting and the land that was once taken from Mexico by the power of wealth is now being repatriated by the power of poverty. North Americans broadcast images of wealth in television commercials and such programs as *Dynasty* and *Dallas*, and the Mexicans respond as multitudes are attracted to the imaginary land of *El Norte*. Neither the flow of electronic information, nor the flow of illegal immigrants recognizes the abstract boundary of the nation-state. The border is not a wall but a very permeable membrane indeed.

But simply eliminating the membrane will not work to solve the problem for either the U.S.A. or Mexico, for the structure of life in a biome is formed around the *difference* between the two regions. In Bateson's terms, it is the difference that makes the difference and constitutes information; but I would take it a step further and say that it is the difference that *drives* the system, that incites to motion. Like a temperature difference that thermodynamically drives an engine, differences stimulate all kinds of human activity, both legal and illegal. The ultimate thermodynamic condition could, of course, become a lukewarm one in which Los Angeles becomes a Third World city and becomes indistinguishable from Mexico, D.F.

Now let us turn from our southern to our northern border. One of the United States's most important exports to Canada is acid rain. In spite of trade barriers and tariffs, and, perhaps, often because of them, North America is a single biome. So, what is integrating the U.S.A.

with Canada? Pollution. What is integrating the United States with Mexico? A movement not of degraded minerals, but economically degraded people; but in both cases of degradation, advertising is the driving force that stimulates a demand of goods that strains the Good.

In both the cases of the southern and northern borders of the U.S.A., the movement is an informational one in which the boundary only serves to establish a difference that energizes and drives the system. Customs and tariffs are differences that structure an economy precisely in the same way in which laws serve to structure black markets and shadow economies. This unconscious organization of life in a biome is not honestly dealt with in the laws that shape a consciously defined polity, for in the cops and robbers game of *Miami Vice*, the interdiction is a subsidy to the crops of the criminals, and the weapons in the hands of the police are permits to murder. In the ambiguity of electronic imagery, the literate and civilized distinction between good guys and bad guys is blurred.

The biome of Latin and Anglo America is, naturally, part of a larger biome, and ultimately the electronic circulation of information in the noosphere and of gases in the atmosphere constitute the single planetary biosphere of Gaia. Such are the politics of life on earth.

At the moment, however, our political thinking and our political systems relate to the past, to the economics and physics of the eighteenth century. As McLuhan once said: "Politicians apply yesterday's solutions to today's problems." This reactionary response is unavoidable and is part of the nature of human perception, for knowledge is, by definition, the organization of the past. Even when we look up at the light of the stars, we do not see the present, but the light of vanished time. What we see as the present is actually the past, and what we sense as the future and write up in fantasy and science fiction is actually the present. Poets, artists, and science fiction writers are not predictors of the future, but, rather, sensitive reporters of the implications of the present. From Hieronymus Bosch, Jules Verne, and H. G. Wells to Doris Lessing, the artist describes what he or she feels and senses, but cannot see. Another word for this mode of perception is Imagination.

What the artist senses, the economist ignores. The economist prefers to describe the modern world-system in terms of territorially based industrial nation-states, but the imagination presents us with the world as a living being whose internal organs are bounded by permeable membranes, and in the life of a membrane, as Lewis Thomas has shown, negation can be a form of emphasis. When we tolerate the presence of aliens within us, we do not always get sick, but when our immunological defenses read thse endosymbionts as aliens in our midst and rise to the task of attacking them, then our defenses trigger the symptoms we know as disease.[2]

Aliens can be viruses, pollen, bacteria, illegal immigrants, or criminals; but in each of these cases, negation can be a form of emphasis that is in unconscious collusion to energize the system it attacks. If marijuana was legalized and its cultivation and distribution were to be taken over by the giant tobacco companies, the shadow economy of the unemployed teenagers in the U.S.A. would collapse, and in many cities there would be conditions leading to riot and insurrection. Organized crime would also lose the structure of its distribution system, and since marijuana can all too easily be homegrown, it is in the interest of the importers and distributors to cooperate with the police in the game of shifting drug habits from marijuana to cocaine.

In this game-structure of cops and robbers, consciousness is merely the *content*, the moral opinions, that enable the *structure* to persist with all its ambiguities intact. Too much consciousness would threaten the continuity of the game, so there is always a tacit consensus that one should not be overly conscious about the evil present in good, and the good present in evil. What the game requires is a simpleminded identification of enemies and bad guys, for the bad guy is the difference that drives the system. Recall that the United States did not pull itself out of the Depression until World War II provided it with a shift to a new defense industries economy, and then consider what would now happen to the American economy if the Soviet Union were to refuse to play the game in accepting the role of the bad guy that energizes the system. Notice that in all three cases of illegal immigration, the drug economy, and the arms race, a simple, linear elimination of a boundary, a law, or a national competition does not correspond to the tacit consensual organization of culture, even though it may correspond to the abstractions of reformers. Culture structures itself around differences that generate organs and organizations; that is to say, that culture organizes itself by energizing differences. Create a condition of sameness, and immediately differentiation among the same will develop, whether we are talking about speakers of a language or members of a single religion or political party.

Now, perhaps, we can begin to appreciate why, though pacifists have been trying since the fifties to eliminate nuclear weapons, they have not succeeded, and that, in fact, the number of thermonuclear missiles has grown. The weakness in liberal thinking is that it focuses on *contents* and not *structures*. If one eliminates an illegal addictive substance by making it legal, one does not eliminate the structure of addiction. If one eliminates a rejection, ones does not eliminate the need for rejection. For example, if one normalizes homosexuality, then the social need to feel oneself outside the tight, bourgeois social system of constraints will search for a new content for the old structure by affirming pederasty or sadomasochism. If one eliminates conventional war, then violence at sports events springs up in its place. If one eliminates the difference between the U.S. and the U.S.S.R. in the arms race, then the engines that drive the scientific systems of both nations stop, and citizens, no longer threatened, no longer vote for the enormous subsidies that Big Science requires.

Here, perhaps, our imaginations can begin to envision what Reagan's *Star Wars* is really all about. S.D.I. is the open consummation of a process that began in the urban revolution of ancient Sumeria. "Civilization" is a misnomer for this transformation; the word should be "militarization," for walls and standing armies are the novel institutions that arose in the shift from neolithic village to literate city. In considering just how much technology has always been wed to militarization, we should not be simpleminded liberals to focus on the content of Star Wars to get caught up in silly debates about whether the anti-ballistic missile system will really work. Of course, it won't work, but the satellites, artificial intelligence systems, and lasers are simply the content; the structure is a planetization and represents the transition from a civilian economy, temporarily mobilized for defense, to a scientific economy permanently organized for research and development. And a scientific economy is one in which serendipity, of finding what you are not looking for, is often its most important spin-off.

All well and bad, but the difficulty in getting a civilian population in a democracy to vote for a scientific economy is that the ordinary citizen is afraid of Big Science; he is afraid of the mandarinism that makes him feel stupid and unneeded. Consequently, the only way to get the citizens of a democracy to vote for the transition to a scientific economy is to frighten them, and then deflect their fears of science onto the scientists of "the enemy," so that our own scientists can enter the picture as rescuing angels of deliverance. To motivate our enemy to scare our citizens, we must, of course, frighten him so that he will fall into the posture that is needed to hold up our economy. So far, the Russians have never let us down. Considering how few Americans have ever been killed by Russians, and considering how many Americans have been killed by Americans in our decaying cities, it is clear that our sense of direction is a little off in matters of defense. Perhaps if we put the reconstruction of the New York subways on the Defense budget, we would be able to find the money rather quickly. In the meantime, I think it is only fair to recognize that the Soviet Union is a close and vital part of the United States. We could lose a few states and still survive, but if we lost the Russians as our enemy, our entire industrial nation-state economy would collapse.

As the replacement of a national civilian economy with a scientific-technical planetization, a lifetime Manhattan Project for scientists, the United States hopes to effect the leap from bourgeois industrial civilization to technocratic planetization without experiencing the discontinuous transition of a "catastrophe." *Star Wars,* like a ride in Disney World's EPCOT, is not an expression of scientific theory and political philosophy, but a projection of fantasy, and for this role Reagan is profoundly expressive of our new electronic body-politic.

President Reagan is the archetypal leader of our post-industrial unconscious polity precisely because he is not a thinker. He is almost entirely unconscious. He is indeed Walt Disney's *Homo ludens* and not Luther, Calvin, or Marx's *Homo faber.* During the rise of the middle classes and the emergence of the bourgeois nation-state, the thinker, and not the military knight or the prince of the church, was the architect of new polities. As Locke was to Jefferson, so Marx was to Lenin; but now in the age of global media, it is no longer the set of the theorist and the pragmatist, but the artist and the actor. As Locke was to Jefferson, so now is Disney to Reagan, for it was Disney who first constructed a media city in which the past became a movie set and the citizen was taken for a ride in fantasies with his own enthusiastic consent. It was Disney who, along with McLuhan, first understood that television would change the consciousness of literate, civilized humanity. Cultural critics like McLuhan and Adorno issued dire warnings about mass deception in the culture industry and the end of Western Civilization, but Disney seemed to have a Kansan American naiveté and trust in the unifying power of popular culture. Indeed with his *Snow White,* Disney himself effected the artistic transition from the folk culture of the Brothers Grimm to pop culture; and to be fair to Disney one must recognize that the transition from oral folk tale to literature is as artistically presumptuous as the transition from literature to film. With EPCOT, however, the shadow side of Disney's mass culture seems more apparent, as if his "Imagineers" now felt that the way to achieve a new political collectivization was not through sad "communist suppression" but happy participation in fantasies of progress. In an age when suburban Christianity no longer had the power of pagan rituals and frightful rites of initiation, Disneyland created frightful rides in which evil was distanced and laughed at, and the past became a visibly comforting artifact in a world that was invisibly hurtling toward a new scientific reorganization of society. The content of Disneyland was the turn-of-the-century small town, but the invisible structure was computerization. The content now of Disney World's EPCOT is the "World of Motion" in which General Motors proclaims the freedom of the individual to go where he chooses, but in the darkness of that ride there is neither choice nor freedom. Similarly, "The American Adventure" of EPCOT brings all the American presidents of history on stage, while two old irreverent writers (Franklin and Twain) obligingly serve as ushers in a memorial service of a civic religion that seeks to give the citizen an uplifting patriotic experience. But all the automaton presidents are controlled by a bank of computers from another place and by a small cadre of scientists and technicians from another time.

And so it is with that other actor and automaton, President Reagan. He is not the real leader, but rather the collective representation of this new unconscious polity. He is not a political thinker and, in fact, all his political thoughts are nostalgic artifacts, decorations, and illusions. Like the plastic fountains and artificial plants in a suburban shopping mall, Reagan's opinions about "Fiscal Responsibility," "Christian Values," and "a Strong National Defense," decorate our unconscious polity in which the shadow economy of drugs, crime, and military spending exceeds the civil economy of the conscious Gross National Product. This economy of violence is now much greater than the old traditional business economy that was founded on the Reformation beliefs of "The Protestant Ethic and the Spirit of Capitalism." Precisely because Reagan is not a thinker he is able to live with these contradictions without being aware of them. "Cognitive dissonance" is not an affliction for this kind of mentality, a mentality that is more of a Durkheimian collective representation than a philosophy. As an expression of the collective unconscious in an electronic, informational society, Reagan has become *the* historical expression of our unconscious polity.

Reagan's *Star Wars* is, therefore, no sudden caprice or casual afterthought, but a deep social and economic expression of the Southern California world view, of that curious cultural mixture of Hollywood fantasies and Big Science. It is neither a thought nor a theory, but an actor's intuition and a sense of timing of what is implicit in the audience and in the audience's historical situation. Sometimes the intuition can sense the outlines of the historical situation more quickly than the intellect, for the intellect can become blinded by mountains of data. President Carter, the nuclear engineer, clearly has a higher I.Q. than President Reagan, but it was precisely Carter's meticulousness that got in his way. Carter approved the MX missile system, a costly behemoth that dwarfed the pyramids as a public works project, but the MX would have only stimulated the cement contractor's business. Reagan's *Star Wars,* by contrast, demands the creation of whole new artificial intelligence systems, fifth generation computers, and an integration of universities and corporations that amounts to a complete transformation of civilian society. Ironically, Eisenhower's Republican nightmare of "the military-industrial complex" has become Republican Reagan's dream.

But that is not the end of it, for there are more than economic implications to S.D.I. In essence, Reagan is challenging the John Foster Dulles concept of massive retaliation. In the darkness of Reagan's unconscious is the dim recognition that thermonuclear weapons are really militarily useless for superpowers, for a nuclear winter prevents their use against continental states, and their scale of destructiveness does not enable superpowers to project their power militarily, to control a sphere of influence, or to stabilize a region of critical resources. Since superpowers can afford large military expenditures, thermonuclear weapons are not attractive investments, precisely because they are costly and useless, and take the funds that could be better invested on the artificial intelligence systems that "smart weapons" require for more surgically precise operations against global terrorism. In the bilateral hegemony in which Reagan telephoned Gorbachev to get permission to hit Qaddafi there is a new historical recognition that though atomic weapons are no longer attractive for the superpowers, they are only too attractive for small nations like Libya, Israel, South Africa, Pakistan, Iraq, and Iran. No doubt Gorbachev's generals encouraged the Soviet leader to give Reagan the go-ahead because they wanted to see if the electronic equipment of the Americans would allow them to fly in the dark, hit Qaddafi's palace, but miss the foreign embassies in the neighborhood. The results were not encouraging for the Soviets, and the prospect of the entire American and Japanese industries set to working on new artificial intelligence systems for *Star Wars* certainly gave Gorbachev something to discuss in Iceland.

Reagan's intransigence in his refusal to give up his commitment to S.D.I. is understandable, for clearly both the U.S. and the U.S.S.R. would like to have some way of removing the threat of punk nations that act in non-European and irrational ways, and, obviously, space is the best place from which to monitor and control hostile flights and launchings; but Reagan's intransigence puts him in the contradictory position of needing to keep the Soviets as an enemy to support the American scientific economy, and at the same time share information so that the Soviets do not drop out of the competition or become a spoiler or punk nation themselves. Since both Three Mile Island and Chernobyl demonstrate that the high technologies of the superpowers cannot be trusted to work without errors, it is clear that neither the U.S.A. nor the U.S.S.R. can feel safe from an accidental misfiring on either side; therefore, anything that makes one side feel threatened enough to move up to a state of Red Alert is to be avoided at all costs. It would appear that the cops and robbers game that the U.S.A. has chosen to play with the Soviets is to challenge the Soviets with *Star Wars* and secretly allow the information to be stolen to insure that the Soviets will not become discouraged to drop out of the competition altogether. And so we can see that "national defense" is indeed an example of negation as a form of unconscious relationship, for the end result of the arms race is a transnational militarization that could be called the U.S.S.S.R.

The political entity that is the transitional form between the industrial nation-state and the planetary Gaia Politique is, unfortunately, the State of Terror. Because we humans are only motivated by fear, fear is what we get in our political groupings. Our ruling politicians terrorize us with thermonuclear Mutual Assured Destruction (M.A.D.), and our aspiring revolutionaries terrorize us with visions of thermonuclear war that they hope will organize the masses behind them; and outside these norms, the completely powerless and stateless terrorize us through terrorism itself. In many ways, terrorism is a form of amateur government; the real professionals in the business of terrorism are, of course, the "legally" constituted nation-states. If one looks back into history to read descriptions of public executions, the pillory, or, for example, William Carleton's descriptions of the hanged men along the roads of ninteenth-century Ireland,[3] one can see that the landscape of terror has been an instrument of governance all throughout the history of "civilization," from the Assyrians to the Aztecs to the British. What has changed is that in an informational, electronic, global polity, a noetic polity, there is no such thing as "space" separating the innocent from the guilty, and so innocent bystanders in airports are chosen to attract public attention and to show that the invisible and stateless, be they Irish, Basque, Palestinian, Corsican, or Armenian, can indeed become visible and able to involve those who thought that they were separate, innocent, and safe into the common lot of death. Ironically, the stateless who long for old-fashioned, concrete, territorial nation-states become media-producers who manipulate electronic, noetic polities through terror.

To effect the transition from the global State of Terror to a more enlightened Gaia Politique, one needs to have a deeper understanding of the unconscious relationships between good and evil, for all too often, terrorists use dim notions of holy war to sanctify their violence in the name of God or the Good. If we ask ourselves, "What is a terrorist and how does he differ from a military revolutionary, be that revolutionary Moses, George Washington, Michael Collins, or Menachem Begin?" we will not be able to answer the question simply in the terms of the motivating or inspiring cause. For the ancient Egyptians, Moses would be seen as a terrorist. For the modern Palestinians, Begin is a terrorist; for the Israelis, Arafat is a terrorist; for the Americans, Qaddafi is a terrorist; for the Libyans, Reagan is a terrorist. If one orders an air strike on a city, and pregnant women are killed, that is consensually seen as war; but if a terrorist chooses to bomb a maternity ward, because he knows that dramatic act will make the evening news, the terrorist act appears to be more calculating, inhuman, and monstrous, even if more people are killed in the conventional air strike on a city. In this dark and irrational world of violence, one becomes lost in a chaos of images and inchoate passions; and in this imagistic and preconceptual world, one sharp television picture of the murder of a single person can appear more horrible than the older cinematic images of a distant city in flames.

If we ask ourselves, "What is Terror?" we will have an even harder time trying to understand this irrational state of human association. Terror is not an object or a place, nor is it an act, like dying. Terror is a state of mind, a condition of complete fusion between perception, knowing, and feeling. Dictators and revolutionaries both seek to create states of terror because they know that in times of chaos and insecurity, they can, paradoxically, comfort the frightened with Terror, and thus achieve the clarity of social fusion through the elimination of ambiguity and confusion.

As a state of consciousess, Terror is a dark and evil foreshadowing of the evolutionary emergence of noetic polities as our next societal organization. Very often in cultural transformation, "evil" is the emergence of the next adaptation.[4] The Viking Terror of the Middle Ages was the first projection outward into the cultural ecology of the Atlantic; it would be followed by waves of explorers, adventurers, privateers, and pirates, before Atlantic, European, industrial civilization would consolidate itself in the more benign forms of post-Enlightenment, middle class, parliamentary democracies.

Buckminster Fuller once observed that the people who first began to think and navigate on a planetary scale were the pirates. The pirates expressed the transition from land-based economies to world trade. The socialist historian of the modern world-system, Immanuel Wallerstein, has even gone so far as to argue that formative capitalism, privateering, and piracy are not as separate and distinct as we have been led to believe by the conventional historians of the British and American nations.[5]

This relationship between good and evil in a world structure seems as much biological as sociological. If one is an anerobic cell, oxygen is evil; if one is a hominid, human tools are evil; if one is a hunter or a nomad, agriculture is evil; if one is a farmer, industrial cities are evil; and now if one is a civilized, literate human being, a noetic ecosystem (or Chardin's Noosphere) is evil: it emerges out of the imagination of science fiction as some kind of demonic noise, psychic vampire, or "Mind Parasite": an *Omni* magazine illustration of flesh embedded in metal and silicon.

Cain the farmer hated Abel the nomadic keeper of sheep. Amos the shepherd hated agriculture and the settled life of cities. Now the other children of Abraham, the Islamic fundamentalists, hate the Israelis and the Americans, the keepers of science and high technology. We have been playing a simple dualistic ping-pong game between Good and Evil since the emergence of writing and civilization, but now that writing is being replaced with electronics, and civilization with planetization, it also seems as if Natural Selection is about to be replaced with genetic engineering. Very little, then, will be left of "nature," and so the nature of Good and Evil is also bound to change. Indeed, all our hysterical fundamentalisms, whether Marxist, Muslim, or Moral Majority, seem *prima facie* evidence that the simple dualistic structure of consciousness is in its final, if deadly, supernova stage.

Now, notice the isomorphisms: noise, pollution, crime, terrorism, piracy, and warfare: all these evil expressions that have their life in the darkness of the human spirit, in the obscurity of the planetary unconscious polity, are similar in that they are formative of structures across the boundaries of old domains, and that existing and traditional domains fight the new in a mode of emphasis through negation. Notice also that in the acceleration of history, the interval in the pendular swing between good and evil shortens as we go higher to the point of fixation. The foundation of the monastery of Lindisfarne is A.D. 635; the first monastery to be destroyed in the opening of the era of the Viking Terror is Lindisfarne in 793. The Renaissance in Florence is 1450; the Inquisition and the burning of Giordano Bruno at the stake is 1600. From late nineteenth-century Wagner to mid-twentieth-century Hitler, or from Disney in the fifties to Reagan in the eighties, the interval between light and shadow lessens as we move upward and inward to the point from which the pendulum is suspended. At that central point which is also at the center of our own consciousness, we can draw our shadow inward and no longer project it outward on some hated enemy, an enemy that conveniently allows us to ignore our own inner capacity for evil. Inside this center, the mythological figures of the evil demiurge and the beneficent Creator are at rest and, presumably, esoterically at peace with one another.

If evil is not a separate condition, act, or being, but a shadow to the process of the emergence of distinct form into the light (just as "death" is the shadow to the process of the emergence of distinct individuals in the evolution of the eukaryotic cell), then one cannot "fight evil" in a simple conflict against conditions, acts, or beings. One has to transform the structure of consciousness and not simply move its contents about in various adversarial positions. When the structure of consciousness is observed (*Vipassana*), and not merely its contents, then we begin to see the codependant origination (*Pratityasamutpadha*) of good and evil, and then we can begin to feel the passing together of all sentient beings in time that is com-passion. Such would be the Gaian politics of an enlightened life on Earth.

From all these separate points one can, perhaps, begin to connect the dots to envision the pattern that connects the disparate elements of the art, science, and religion of our contemporary culture. Noise, pollution, crime, terrorism, and warfare all constitute unenlightened and unconscious forms of activity in which we say one thing and do another, in which we *are* one thing but *act* another. In each case we seek to expel an alien into an imagined external space that will permit us to continue our conscious agenda unchallenged by responsibility to the entire pattern of biological relatedness. But if noise is attended to, if pollution is transformed into a re-source, and if enemies are seen to be intimate projections of our own internal life, then our political systems will begin to change. They will move away from the modern mechanistic descriptions that shape our economic conceptions, and they will move away from the medieval religious notions that shape our holy wars of Jew against Moslem, Sikh against Hindu, Protestant against Catholic. As we move beyond the religious sensibility of the traditional medieval civilizations, and beyond the physical sciences of European modernism, we will begin to appreciate the organizations of the living expressed in biology and ecology. Both Darwinism and Social Darwinism are old forms of thought in which competition for survival occurs within a limited space. "Space" no longer has the meaning it did for our Victorian ancestors, and some newer forms of biological thinking, such as Lynn Margulis's *Symbiosis and Cell Evolution*[6] would suggest that cooperation has as much to do with evolution as competition; but the application of biology to economics that is, perhaps, the most appropriate to our era of cognitive science is expressed in the new cognitive biology of Maturana and Varela.[7]

II. Toward an Autopoetic Economy

An application of the theories of "autopoesis and cognition" to politics and economics may seem a little farfetched, but to begin to appreciate just how the theories of Maturana and Varela could relate to the economy of an electronic era, we need only consider those seemingly irrelevant areas of pop culture that conventional industrial economists ignore. For example, let us begin with the punks on the King's Road in Chelsea, London. The punks are an industrial proletariat that has recycled itself into an informational proletariat. Knowing full well that they were not needed by the upper and middle classes, not as slaves, serfs, or factory workers, they did not wait for the monetarists of Thatcherdom to tell them what to do with their lives, but entirely on their own they went on to invent a life-style that spins off its own economy. And that is as good an example (better, from a moral point of view) of an autopoetic economy as Reagan's *Star Wars.* This way of life as art-style spins off a music industry, a fashion industry, and these, in turn, spin off a music video industry and a whole series of associated magazines and newspapers. A new informational middle class begins to sprout up in the media, and this middle class begins to live off the creative energy and innovations of the lower class. Now, if one adds up the entire sum of all these transactions on a global scale, and then one divides this sum by "the dole" that the punks received as "unemployed members of the working class," one will begin to see what a tremendous return on investment the dole represents. Perhaps the old notion of a guaranteed annual income would not be quite "the drain on the economy" that industrial-age economists project.

By way of contrast, and in the British spirit of fair play, consider all the money earned by nuclear power plants and the Concorde, and then divide that sum by "the dole" given to the managerial class "to stimulate the economy" by subsidizing nuclear energy and the development and maintenance of the Concorde. Very few people are affected by the Concorde, but hundreds of millions of people are affected by the music industry, even the starving in Africa. And yet, in spite of all this economic and cultural enterprise, and, ironically, in spite of everything the punks and rockers do to become highly visible, Thatcher and her cronies cannot see them as anything but noise. In dismissing them from her consciousness, Thatcher puts her attention on the "real" economic business of rescuing an insignificant helicopter company. Clearly, the perception of reality has little to do with economics; rather, economics has a great deal to do with mythically shaped modes of perception. As we begin to appreciate the fact that an economy is not only based on gold and natural resources, but also on culture, and that one of the reasons why the Swiss and the Japanese are prosperous is because they are Swiss and Japanese, we will begin to understand that music, very much like an economy, is a global noetic ecosystem. In fact, music may very well be the polity of the future.

Today, 80 trillion dollars flows around the world daily, but only 15% of it is used for physical trade;[8] the rest is involved in an informational flow, as, for example, in the arbitrage of currencies, in which *difference* drives the system. As we begin to appreciate this newer kind of "groundless" cognitive science,[9] we will accelerate our transition from an industrial to an autopoetic economy.

This illustration of economic development taken from popular music instead of "real" industries like coal or railroads, or oil and aerospace, is not as fanciful as it may seem. The economic adviser to France's President Mitterrand, Dr. Jacques Attali, has argued in his *Noise: The Political Economy of Music* that the developments of western music anticipate the social developments that later become consolidated into economies:

> THE FOUR NETWORKS
>
> The first network is that of *sacrificial ritual,* already described. It is the distributive network for all of the orders, myths, and religious, social

or economic relations of symbolic societies. It is centralized on the level of ideology and decentralized on the economic level.

A new network of music emerges with *representation.* Music becomes a spectacle attended at specific places: concert halls, the closed space of the simulacrum of ritual—a confinement made necessary by the collection of entrance fees.

In this network, the value of music is its use-value as spectacle. This new value simulates and replaces the sacrificial value of music in the preceding network. Performers and actors are producers of a special kind who are paid in money by the spectators. We will see that this network characterizes the entire economy of competitive capitalism, the primitive mode of capitalism.

The third network, that of *repetition,* appears at the end of the nineteenth century with the advent of recording. This technology, conceived as a way of storing representation, created in fifty years time, with the phonograph record, a new organizational network for the economy of music. In this network, each spectator has a solitary relation with a material object; the consumption of music is individualized, a simulacrum of ritual sacrifice, a blind spectacle. The network is no longer a form of sociality, an opportunity for spectators to meet and communicate, but rather a tool making the individualized stock-piling of music possible on a huge scale. Here again, the new network first appears in music as the herald of a new age in the organization of capitalism, that of repetitive mass production of all social relations.

Finally, we can envision one last network, beyond exchange, in which music could be

> lived as *composition*, in other words, in which it would be performed for the musician's own enjoyment, a self-communication, with no other goal than his own pleasure, as something fundamentally outside all communication, as self-transcendence, a solitary, egotistical, non-commercial act.[10]

In the first network of ritual, we should envision peasants, panpipes, and music played out-of-doors at some harvest festival. In the second network of representation, we should envision Mozart playing before the Archbishop of Salzburg. In the third network of repetition, we should envision Toscanini in the recording studio, laboring over the symphonies of Beethoven. And in the fourth network of *composition*, or what I prefer to call autopoesis, we should envision two teenagers, one in Los Angeles, the other in Sydney, "jamming together" by playing self-composed, computer-animated music-videos through the use of their own personal computers, modems with satellite hook-up, and VCRs. This composition would be improvisational jazz and imagery, a *poperatic* art form that could melt away, or if the teenagers preferred, could be saved on video discs. An electronic conferencing of a group of these teenagers would constitute a nightclub that was not "simply located" in space or time. In other words, two teenagers with personal computers and a VCR would be able to create *poperatic* art forms that for a generation before would have taken an entire recording and television studio. If we expand our imaginations to envision a jam session of teenagers all over the world, we will be able to see the emergence of a global noetic polity. Such is planetary culture.

Although Dr. Attali was an economic adviser to Mitterrand's socialist regime, his theories work much better for entrepreneurial and electronic California than they do for literary and bureaucratic France, which is one reason, perhaps, why Dr. Attali has such difficulty in imagining the fourth economic network on anything but negative terms such as anti-social narcissism. Very strangely, Dr. Attali does not seem to be aware that his four networks constitute a fourfold Viconian cycle; were he more familiar with Vico, he might begin to see that in a *corso-ricorso*, the fourth network has many of the characteristics of the first. Thus in an autopoetic economy of global music, the planetary nightclub and the planetary gathering take on qualities of individual and collective at the same time. This is not a case of print-isolated man alone in his book-lined study, but McLuhan's global villager at an electronic, rather than an agricultural, festival.

An autopoetic economy is, by definition, polycentric and popular. French culture, by contrast, stresses centralization in Paris and conformity to standards of language, thought, fashion, and behavior, set by a small elite and held in place by a rigidly conformist bureaucracy of educators, civil servants, and waiters. This ethos serves well for grammar and cuisine, but it does not serve to inspire a popular, and, indeed, vulgar culture of irrepressible innovation, risk taking, and educational liveliness. Because French culture is so hermetically sealed unto itself, all the mental action is an internal digestion process by a group of anaerobic intellectuals. This sort of culture can produce good cheeses and wines, but it does not produce the banality necessary for an exhaustive atmosphere of innovation and noise. It is small wonder that the French are upset that English has become the language of the informational society, and that California has outstripped France in fashion and science. The minimalist English of American engineers has eliminated all sense of nuance and style as memos replace essays, nouns are impacted into verbs, and subtle grammatical distinctions, such as those between "as" and "like," go the way of the old British archaisms, "whither" and "whence." Against this electronic flood of tech-talk, the seventeenth-century elegance of French (as well as the medieval weightiness of German freighted down with all its clumsy packing cases) is as hopelessly out of place as a Deux Chevaux in orbit.

The economic failures of leftist Mitterrand and the economic successes of right-wing Reagan hold out many lessons for us intellectuals. Although Mitterrand's I.Q. is no doubt higher than Reagan's, the two politicians are alike in that their ideologies have little to do with their political behavior. Mitterand, though nominally a socialist, is no different from De Gaulle when it comes to hauteur, delusions of French grandeur, and a love for thermonuclear weapons. Reagan, for all his fundamentalist, Moral Majority rhetoric, is no Christian Ayatollah, and he has done more to consolidate the shift from industrial to postindustrial than any president since Kennedy. Reagan's peculiar talent seems to come from the creative ability to entertain opposites, and perhaps this comes about because he has so little consciousness to get in his way. These apparent contradictions, however, are not so novel as one might think. Liberal, industrial England was ruled by the conservative collective representation that was Queen Victoria, and Nazi Germany was led by a nativistic leader who created an untraditional merger of technology and the state.

If we go farther back into prehistory, we can see that this pattern of innovation disguised as conservatism has always been there. For example, in the period of the neolithic revolution, when humanity, thanks to women, was effecting the shift from hunting and gathering to agriculture, the iconography on the walls of Anatolian Catal Hüyük (6500 B.C.) celebrated hunting. The economic structure of the culture was neolithic and agricultural, but the content was paleolithic. Similarly, when we look at the shift from medievalism to modernism, we see the same archaistic features. In Renaissance Florence, Cosimo de Medici is caught up in visions of Plato's academy and neoplatonic mysteries, but the structure of the new culture is based on new forms of communication in banking and art. And when we consider the next major historical shift, the shift from agricultural to industrial society, we find the same pattern: the structure of the Great Exhibition in London in 1851 is industrial wrought iron and glass, but the content is medievalism and romanticism.

It goes without saying that Reagan is no cultural historian and has not modeled his behavior upon a study of neolithic Anatolia, Renaissance Florence, or nineteenth-century England; so, we must assume that whatever it is in human culture that produces this pattern is precisely the same collective unconscious force that throws up Ronald Reagan onto the screen of history. In the literal Latin sense of the word, Ronald Reagan has a political *genius*, or, at least, he is an idiot savant. On "A Gala Night of Stars," one more reminiscent of Las Vegas than New York, Ronald Reagan relit the Statue of Liberty with a laser. In the company of his Hollywood colleagues, Reagan gave America so clear a performance of its collective representation that it is now obvious in his second term that Reagan's use of the media to collectivize society is even more skillful than F.D.R.'s use of the radio in the Depression. F.D.R. used the radio, Hitler used the loudspeaker, and the Ayatollah Khomeini used the tape cassette recorder, but no politician has ever used television so cunningly. Now even Reagan's opponents have to admit that he is not an actor become amateur politician; rather, it is the old style politician who is the amateur public personality.

If, as I have argued above, that *Star Wars* is analogous to a display in EPCOT, and is, therefore, an illusion and a piece of pure show business, it is none the less serious business. Show business is not based on "real goods" but on created desires. Advertising wed to public credit seems to have been part of the reality of postindustrial society in the 1950s, and now show business wed to Big Science and national credit seems to be part of the reality of Reagan's post-postindustrial society. In a global economy with an 80 trillion dollar flow, it is difficult to say that such a movement is backed up by anything, land or gold. Currencies are good not because they are backed up by goods, but because people believe in the viability of that nation. Nations are now marketed and judged the way companies once were. Considering that such gaint multinationals as Shell and Nestlé have transferred billions of investment from the Third World into America, and considering that America is now a debtor nation no longer "owned" by its citizens, it does seem as if the investors are saying that they believe that the U.S.A. is going to make the transition from civilization to planetization, and that in this quantum leap up to a new level, perhaps one even greater than the shift from agricultural to industrial, they do not want to be left behind in the nativistic position of an anti-American Iran. The fact that European, Middle Eastern, and Asian capital is flowing into the United States means that some people are betting their money that the U.S.A. with its Harvards, Berkeleys, and Stanfords, its M.I.T.s and Cal Techs, is going to make it.

England borrowed from the Netherlands to build the railroads and canals that made its eighteenth-century industrial revolution possible, and then the United States borrowed from England to finance its nineteenth-century industrial revolution. Now through treasury notes and huge deficits, the United States is borrowing from the world to finance its transition from a civil to a scientific economy. Since no nation ever seems to pay back its debts, but merely pays the debt service that allows the game to continue, it is not at all clear just what it means for a "nation" to be in debt in a planetary culture. Perhaps the definitions of ownership and sovereignty are no longer appropriate for this historically novel situation.

Both Disneyland and Disney World openly celebrate fantasy and play, values certainly no Moral Majority figure such as Calvin could countenance. For Calvin and Luther, the world was no laughing matter. For "The Gutenberg Galaxy" as a whole, the world was earnestly re-presented in the Bible, in print, and in written laws. But for *Homo ludens, representation* gives way to *participation.* This shift is not simply a change in ideologies, but a systemic reorganization of human culture. For Locke, ideas were mental representations of the outside world: and for Locke as well Parliament was the brain of the body-politic; it was filled with representatives of the society at large. Underlying both the theory of knowledge and the theory of the state was a literate consciousness in which the wild "commons" could be turned into "property" in much the same way as the "commons" of oral speech could be turned into written language. Walls, fences, and lines of print have indeed much in common in deeds and commons.

We can see all these connections of modernism now because we are in the process of leaving modernism behind. In post-modernism, the brain does not represent the world, and the member of Parliament does not represent the body-politic at large. The isolated citizen represented by a politician is exchanged for the participatory individual living as a symbiotic organelle empowered by information in an environment that is not structured by institutions such as church and state. A good example of this political shift is expressed in Greenpeace's ability to challenge the nation-state of France. Considering the power of electronic modes of communication in transforming culture, it is fair to say that there is more of a future for political entities like Greenpeace, Amnesty International, or Africa Live-Aid, than for industrial nation-states trying to extend nineteenth-century patterns of imperial domination from Europe into the South Pacific.

In the representationalist paradigm of Lockeian philosophy, the mind carries a little picture of the world inside its head, parliament carries a little picture of the outside polity within its chambers, and paper money carries a little picture of national gold and land within its rectangular form. In the cognitive science of autopoesis, however, the brain is no longer seen as the house of representatives. The economic analogue of this autopoetic paradigm is that money is no longer seen as standing for reality. Money is no longer backed up by national land or gold, but by the belief in a nation's productive capacity for scientific innovation. The autopoetic economy creates its own values in transactions. Since cultural transitions such as the one from hunting and gathering to agriculture, or from agriculture to industry, are so unpredictable, the behavior of an autopoetic economy is without precedent and takes on the quality of a self-fulfilling prophecy. If the debtor nations default, and if the world at large begins to believe that the United States cannot make the transition to a new planetary culture, the U.S. won't. Part of the Gaian politics for the nineties is, therefore, to realize that we are all organelles within a planetary cell, and that it is a dangerous illusion to think that any nation-state can make it on its own, militarily or economically.

The interpenetration of all in each argues that territorial sovereignty is a leftover from the representationalist paradigm of the seventeenth century of Locke. Greenpeace can intrude into the politics of France, and Nestlé can buy up Carnation in the U.S.A., Chernobyl can ruin the agricultural produce of Eastern Europe, and middle-class American college students on cocaine can sustain the Shining Path along the Andes. Voting for one's local representative in congress or parliament does not give one a handle to the door of this new world; it merely shuts one out, and the citizen knows this, and that is why participatory groups, from Greenpeace to Live-Aid, are so much more charismatic than elections in our global electronic culture.

An image of David challenging the Goliath of the old industrial nation-state should not mislead us into thinking that there is only a positive side to the participatory mode of politics. There is definitely a shadow-side to the shift from representation to participation. Representation expresses the culture of a civilized consensus. In the time of Thomas Jefferson, a civilized man had only to read one or two hundred books to be educated; now a hundred books is what appears in a single specialty within a year. In this world of information overload, the benumbed citizen no longer reads or thinks; he watches and feels. The kind of individual who is uplifted by the civic religious propaganda of EPCOT is not your literate Benjamin Franklin or Mark Twain. Spectator participation in media-fantasies is a return to the peasant's illiterate participation in the medieval pageantry of knighthood and Church festivals, and is clearly not the same kind of involvement as expressed in personal middle-class power and elected representation. As in the case of a rock festival or discotheque for the hipper crowd of the young, EPCOT has been designed so that noise becomes the solvent that breaks down the isolated individual and bonds him to the group. With loudspeakers in the flower beds blasting the pedestrian with majestic theme music from dramatic moments in motion picture history, the citizen becomes a technopeasant stripped of his right to silence and private thoughts as he is bound and rewound into the taped continuum of musak.

But it is also in this world of noise, global communication, and individual alienation, that angrier souls can use their modems and electronic bulletin boards to form racist groups such as Aryan Nation. It is in this world of fragmentation that fundamentalism and terrorism both seek to melt the bits into molten lead, for, unfortunately, terrorism is also an extremist form of *participation* replacing *representation* in electronic politics.

Undoubtedly, nativistic movements that are simplistic and violently reactionary will continue, even in the U.S.A., but those in the U.S.A. will most likely fail. As we have seen from American television, TV evangelists are no different from rock stars or Hollywood celebrities, and as the fundamentalist flips from channel to channel, changing realities from Westerns to science fiction, from soap operas to news, from one part of the world to the other, he is like a visitor moving through the rides of Disney World and he is participating in and performing a sensibility that is radically different from that of his Reformation forebears. The only way that Reformation culture could maintain the "Gutenberg Galaxy" is to eliminate television in an Amish-like purity of freezing history at an earlier moment of time. And probably some fundamentalists will try to go back when they begin to wise up to the fact that Reagan is a liberal and that evangelists on television are not ministers of The Book, but ministers of the media.

The U.S.A. is, of course, too far gone, thank God, to turn back and lock itself into an Aryan Nation. The U.S.A. is more addicted to TV than cocaine, and its continental polity has become even more multiracial than it was in the nineteenth century. Now it is no longer simply a case of integrating Italians, Irish, Eastern European Jews, and Blacks, but Mexicans, Cubans, Vietnamese, Koreans, and Japanese. As New York was the quintessential city of the U.S.A.'s nineteenth and early twentieth-century expansion, so now Los Angeles is the quintessential world city of its electronic culture. In the earlier print culture, the public school teacher could take the child away from its parent's old world culture to make it all-American, but in the multi-channel culture of television, ethnicity is reaffirmed and immigrants keep their languages, Spanish or Korean, and, in fact, Americans begin to develop new ethnic languages like Jive. The world of electronics is Top and Pop culture; it is an energizing of opposites: the elitist science of Stanford, Cal Tech, and the Silicon Valley, and the media and musical reproductions of Hollywood.

This structure of opposites is economically very important, for one of the reasons so much of the world's capital is flowing into the United States to finance the planetization of humanity is precisely because the U.S.A. is so multiracial and innovative in many different aspects of human culture.

The cultural symbiosis of Top and Pop is the secret of America's present strength, and this strength will most likely continue into the twenty-first century. Japan is not going to replace the U.S.A. as a world power, for Japan is an island culture, and as such lacks the diversity and imaginative daring to take on the role of world leadership. It is strong precisely because after losing the war it immediately chose to become an organelle within the cell of the American world economy, and in that position it could flourish and thrive. For Japan to try to reverse the situation and make the U.S.A. an organelle within the cell of Japan Inc. would lead to a massive stretching and distortion that would simply tear Japan apart. Its present position is strong precisely because its present position is its strength.

If the U.S.A., on the other hand, is to continue to grow in the shift from civilization to planetization, it will have to relate to China as well as Japan and Korea, and in this Pacific world the white-racist mentality of the nineteenth century and early twentieth century is dead wrong. American business, from Nixon to Reagan, knows this and has already decided on the Pacific shift in the economic and cultural centers of America. It is only the passed over and uneducated "poor white trash" who feel threatened by the dynamism of the Asian peoples in the universities of the West who are attracted to nativistic movements like the Aryan Nation and Posse Comitatus—just as before the dispossessed whites who were left behind by the Industrial Revolution that swept over the North after the Civil War were similarly attracted to the Ku Klux Klan. The only way that the U.S.A. could become an Aryan Nation is if it collapsed economically, chased out all its unemployed multiracial scientists, dissenting artists, and intellectuals into whatever country would welcome them, Australia, Canada, or Brazil, to implode in on itself in the simplistic, Biblical, neo-Nazi fascism envisioned by the Canadian novelist Margaret Atwood. In the historical long view, Ronald Reagan, like an innoculation that gives us a little of the disease to save us from a lot of it, may just have saved the U.S.A. from white racist fascism by playing the role of right wing American President for the Moral Majority.

If the United States is going to continue in the transition from civilization to planetization, then it will need to come to a more ecological understanding of the interaction of differences and opposites within an emergent domain. If noise holds the unrecognized signal of innovation, if evil is the emergence of the next adaptation, if the arms race with the Soviet Union is creating the U.S.S.S.R., if the drug traffic is integrating in the shadows the debtor economies of the U.S.A. and its Latin neighbors, and if pollution is bringing us all under a cloud that moves without respect to the boundaries of nation-states, then what is the role of the Good in this cultural transformation? Is only Evil all-powerful, and is the Good always inane and impotent?

The point of the ancient Buddhists was, of course, that Good-Evil are in codependent origination, and that Good is not absolute and transcendent. Those who cannot see the evil and unenlightenment in themselves, in their own conditions of suffering and samsara, project it outwards and inflict their goodness on others, turning a disconnected and uncompassionate virtue into an abstract kind of cruelty. This is the mode of the extreme moralist, a Rabbi Kahane, an Ian Paisley, an Ayatollah Khomeini. The person who feels compassion can sense his own internal capacity for evil, and therefore has compassion for others, even those who have been momentarily taken over by evil. In other words, the good Buddhist is a good Christian and loves his enemy as Jesus counseled.

One needs another word besides Good for this balance at the Center. Christians would call it love or *caritas*, Buddhists would call it compassion, or *karuna*, but in this new balance at the Center, the world is essentially the same and quintessentially transformed. As in a move in the martial arts of Akido or Tai Chi, all the inertial mass of the evil opponent is deflected, and by the slightest and subtlest of moves, the enemy is sent to the ground to find his own center.

What would be the slightest and subtlest of moves that could transform our political world from the global noetic polity of the State of Terror to the planetary noetic polity of compassion, of com-passion or passing with others through the "catastrophe" that is the discontinuous transition from one world system to another?

The martial art would teach us to take what is given, and slightly redirect its energies; to take the form in front of us, instead of waiting for the Messiah or the end of the world, and transform it. So, let us consider the political forms of activity that are in front of us, and imagine what their slight transforms might be.

PRESENT FORMS	FUTURE TRANSFORMS
Star Wars.	A Transnational Program for the Exploration of Space.
Pershing II's and Cruise Missiles to defend Germany.	Withdrawal of all Pershing II's, Cruise Missiles, and atomic weapons from West Germany and the transformation of the *Wehrmacht* into a Swiss-style civil militia.
The United Nations as a failed world government and global police force.	The UN as a world Harvard, a world Academy of Arts and Sciences serving as the third house in a Tricameral legislature in which the nation-states have upper and lower houses, Lords and Commons,

PRESENT FORMS	FUTURE TRANSFORMS
	Senate and Congress, but that all nations have the UN as their third house to provide research and recommendations for such long-term problems of human civilization as the Greenhouse Effect, acid rain, drug traffic, human rights, etc.
International Monetary Fund as a "device for taking money from the poor in rich countries and giving it to the rich in poor countries."	Establishment of planetary Land Grant Colleges, "Gaian Colleges," as Bioregional Resource Centers to set up the beginnings of an informational economy in impoverished areas like Chad or Haiti.
Subsidies to selected institutions or favored groups, such as nuclear industry, the oil business, farmers.	An American Expression Card, or direct venture capital to each citizen in lieu of guaranteed annual income: a sum of $50,000 is granted to each individual at age of majority for starting a business, subsidizing a college education, or letting money earn interest until citizen decides upon a personal investment. Citizens who did not feel competent to invest could leave the money in what would be, in essence, a national mutual fund.

The arguments in support of PRESENT FORMS fill up the media, so rather than waste time on them I would prefer to give my reasons for the proposals under FUTURE TRANSFORMS.

Anyone alive in this historical moment faces societies that are totally structured upon warfare. The governments, the communications systems, the sources for technological innovation, and the entire economy of the United States is held up by the arms race. Any pacifist who comes along and says, "Disarm!" hasn't, literally, a chance in hell. Some pacifists love projects that fail because it gives them a sense of sanctification in a fallen world; other pacifists are simply violently aggressive personalities who cloak their aggressions by screaming for peace. This is the kind of pacifist who thinks he is helping the peace movement by throwing blood onto a military officer. Such forms of opposition only confirm the opponents in their mutual positions, and thus energize the game of pacifism versus militarism, a game as useless as that of wealth celebrating the virtues of poverty from the throne of the Papacy.

Confronted with a war economy, one has to build down slowly and shift employment to other areas. The difficulty is that citizens and politicians will only vote for subsidies under threat, and so there always has to be a threat from the enemy or the environment to mobilize a society. But if people really begin to feel threatened by the warfare system, if they now begin to see, after Chernobyl, that we cannot trust the Russians to man their nuclear deterrent with a fail-safe system, then, perhaps, the citizens will begin to understand the need to have a vigorous space program as a way of keeping the "Star Wars " economy going with more stars and less war.

In his second Presidential Debate against Mondale, Reagan proposed sharing Star Wars with the Russians. His generous flourish of "Why not share it with the Russians?" was a brilliant rhetorical move, for it made him into the liberal and Mondale into the cold warrior conservative, a position that Mondale could not convincingly maintain, since he could hardly upstage Reagan on talking tough to the Russkies. It was precisely these kinds of maneuvers that enabled Reagan to steal the future as political mythology away from the Liberals and emerge as the new champion of futurology, scientific innovation, and the Space Age: which, considering Reagan's level of reading and scientific literacy, is no mean feat.

The difficulty is, of course, that Reagan was playacting. We already have a satellite reconnaissance technology years ahead of the Russians, but we do not seem willing to share that with them now. So if we won't share now while we are ahead, why will we share later?

Still, that uncanny intuition of Reagan was right, even though his political colleagues would never want to follow him. We should share with the Russians, but *now*, not later. Anything that makes the Russians feel paranoid and insecure threatens our security. If we get close to building a real ABM system, the Russians would have to launch a preemptive strike before we completed it. Of course, Reagan's real strategy is to force the Russians into an expensive arms race in the hope that their strained economy would collapse before ours would. But that too is dangerous, considering the fragile condition of the world economy. With our two trillion dollar indebtedness, it is a high school game of economic chicken to see who collapses first, we or the Russians.

But let us assume our media leader is right: we launch into Star Wars and create a new third industrial revolution that puts the Soviet Union back into the status of a Third World, underdeveloped country, and we interlock all the economies of Western Europe and Japan into "Trilateral" subcontractors for the new Space Age. We should recall that it was a humiliated and defeated Germany after Versailles that led to the rise of Hitler. Old Russia, no longer endowed with the facesaving position of being the equal of the U.S.A., a mythic identity bestowed on it by Nixon and Kissinger in the Breznev era, would have no choice but to play the spoiler. Since we have seen what kind of chaos spoilers with ravaged pride and identity can do, whether they be Palestinians, Libyans, or Iranians, it would be wise not to push the Soviet Union down. It would be safer for all to start sharing the satellite system of reconnaissance immediately, so that everyone knows what everyone else is up to.

The human gesture of shaking hands was invented as a way of showing that the hand did not conceal a weapon. It is a nice custom, one that was raised on high when the Cosmonauts and the Astronauts linked up in outer space and shook hands. We need to go back to that point and take it from there.

Rather than surreptitiously arranging for the Russians to steal our technological secrets, we should simply let them try to compete with Germany and Japan for contracts. When their technology is not good enough, then that will encourage them to find out, in the new Management Schools of Gorbachev, why their system of production works well for some things but not for others.

An immediate sharing of satellite reconnaissance will help all nations have a more effective defense system, thus lowering budgets, and making resources available for other schemes of development than having every poor nation in the world trying to buy American or French jet fighters to defend their borders. Since some of these satellites can now be marketed by corporations, it also means that civic groups such as Greenpeace could have their own systems of watching who is dumping atomic wastes into the seas. Such civilian surveillance would also make it hard for one nation to launch a sneak attack on another, as in the cases of Iraq and Iran, and this would force nations to come to terms with World law and the World Court.

If expensive forms of mutual defense are replaced by simpler forms of mutual security, then the whole World War II structure of the world system can, finally, change.

Indeed, the best way to avoid World War III is to stop fighting World War II, and this brings me to the point of my second Future Transform.

Part of the condition of the global State of Terror is a kind of thinking in which "Mutual Assured Destruction" or MAD constitutes a sound policy for national defense. Since no one is literally defended in such a situation, the result is simply that national populations are held hostage by military-technical elites: the really professional terrorists of the world. If shared satellite reconnaissance would enable the Russians to feel less threatened behind their "sacred borders," then Europe could continue to contribute to Russian security, and thereby to mutual security, by not having the offensive capacity of decapitating the Soviet Union within minutes through the stationing of Pershing II's in Western Europe. If the German *Wehrmacht* were converted to a Swiss-style civilian militia, it would be possible to *defend* Western Germany without a military force that was *offensively* capable of destroying the Soviet Union. If the American troops were to leave Germany, as a clear expression of ending World War II, then the Warsaw pact countries would begin to have increased trade relations in the world economy, and the demands for liberalization, already pronounced in Hungary, would begin to be characteristic for Eastern Europe as a whole. What helps the Soviet Union keep Eastern Europe under tight control is the nuclear threat from the West. If this is dropped, then Eastern Europe will begin to evolve culturally along new lines that are no longer the outlines of the Cold War of the fifties.

If Western Germany were to finance its own civilian militia, the deficit in the U.S.A. would be reduced, for NATO now consumes half the military budget, but this massive investment is not really militarily effective, for the 300,000 U.S. troops in Europe could never repel a serious Soviet invasion, and it is highly unlikely that the U.S.A. would risk either a nuclear attack or a nuclear winter by defending Western Europe with nuclear weapons. Since the bulk of America's trade is now with the Pacific Rim countries, the U.S.A.'s attention is much more focused on Asia and Latin America than Europe. Clearly, Europe is going to have to look after itself, and in this new position, Switzerland, and not simply Finland, is a model for Norway and Western Germany.

If Western Europe begins to participate vigorously in a Transnational Space Program, then these planetary forms of knowledge will give a new historical role for the United Nations. Now the United Nations is a postwar creation that is as out-of-date as NATO. It has never been able to stop a war and has been a miserable failure as a vision of men beating their swords into ploughshares, but if it were allowed to be truly a world cultural organization and not a political organization manqué, it could mean more than a Manhattan life-style for global bureaucrats.

In the cultural evolution of civilization from print to electronics, societies need to shift from the bicameral legislatures of the days of Locke and Jefferson to tricameral legislatures. The notion of a bicameral legislature is one of balancing urgency with reflection. The lower house is intended to respond to the needs of the moment, the Senate or the House of Lords is supposed to express age, wisdom, and the longer civilizational viewpoint. In a scientific, planetary civilization, however, the longer viewpoint requires more than a gentleman's knowledge of history and the classics. When confronted with problems like the greenhouse effect, acid rain, the poisoning of the oceans, global drug traffic, world terrorism of transportation, and human rights in general, knowledge is needed along with wisdom. No nation is ever likely to surrender sovereignty willingly to a world state, and a good case can be made that a world state would be at once bureaucratic, ineffective, and tyrannical. But every nation does surrender to the information that comes to it from global forms of communication, world science, and popular art forms like music and movies. If the United Nations were to become a global Harvard-M.I.T. with bioregional resource study campuses spread around the world but communicating through electronic networks, it could become a planetary Academy of Arts and Sciences that could make reports and recommendations to the bicameral legislatures of the nation-states. Some nations, like Switzerland, might be quick to implement, for example, speed limits and air pollution controls to stop the death of the forests; other nations, like Italy or Germany, might be much slower to respond to the need to rein in its drivers, but the successful action of smaller and more effective regions of local control would be highly instructive to others.

As the third house of every nation's form of self-governance, the United Nations could become valuable precisely because it was not a direct form of government. No one questions the valuable economic contributions of an M.I.T. to Massachusetts or of a Stanford to California, so it should be possible to envision a global informational culture in which such a planetary body would be vital to local economies and governments. If horribly impoverished countries like Chad and Haiti had bioregional resource centers doing work on desertification, soil loss, and medical research, these small colleges could become the equivalents of the Amhersts and Pomonas of the U.S.A.: small, traditional liberal arts colleges doing well in an age dominated by giants like M.I.T. and Stanford.

These small planetary Gaian Colleges could become the twenty-first century's equivalent of the land grant colleges that the United States set up in the nineteenth century to develop the country. One of these land grant colleges became Cornell University, which is now a wealth-producing, and not merely a wealth-consuming, world university. If the United States and the Soviet Union were to agree to build down their ICBMs, one Trident submarine would more than pay for several of these Gaian Colleges.

It is important to keep in mind that when I say "college," I don't mean simply Ph.D.'s from Harvard on salary in Chad or Haiti. I envision something much more like the Land Institute in Salina, Kansas or the Meadowcreek Project in Fox, Arkansas—institutes in which local people are colleagues and not merely subjects of research for national elites.

The former postwar development scheme was simply the Americanization of the planet. Loans would be given so that markets for American goods and banks would be created around the world. Dams would be built, jet planes would be sold to dictators, and village agriculture would be bulldozed so that American agribusiness could establish itself. Now forty years after this 1946 vision of world progress, it is time to say that this scheme for development has been a miserable failure. The rich have gotten richer, the poor have gotten poorer, and the middle class has been eliminated in the cross-fire between fascist repression and communist liberation. This scheme of development has been good only for the makers of jet fighters, machine guns, dams, and nuclear reactors.

But the lessons to be learned from the failure of these schemes of Third World development can also be brought home to the so-called developed countries. Development schemes have not worked abroad, and development schemes have not worked within the U.S.A. domestically. The failure has come from trying to give only to *favored* groups or institutions, whether through oil-depletion allowances, farmers' subsidies, or the complete underwriting of entire industries, as with the case of nuclear power. A shift in paradigms would call for providing venture capital directly to *each* citizen as part of his *participation* in the national economic community. If each citizen were to be granted upon reaching the age of eighteen a venture capital of $50,000, it would do more to stimulate the economy than subsidizing nuclear power or the oil business. If two California teenagers in a garage can start Apple Computer Company, and if teenagers at large can create vast markets and industries in music and video, it is clear that the archaic industrial mentality that subsidizes behemoths like nuclear reactors, but balks at giving money directly to its citizens is simply incompetent business. The new technologies now make it more possible than ever for a group of teenagers to band together and create their *poperatic* works of art and business. Of course, some teenagers will use the money to go into the drug business, but with a direct venture capital fund available to them, the economic motivation to be drawn into the shadow economy will be lessened, and kids in ghettos will be given a choice that they now do not have. For some, this may mean using the funds to finance a university education; for others, it may mean starting a business, and for others who are not ready, it may simply mean letting the money stay put to earn interest in what would be, in essence, a citizen's national mutual fund. This "American Expression Card" would be the citizen's patrimony and the visible sign of his participation in the autopoetic economy in which all invest in each, and each creates new economic worlds for all.

This pacific shift of Star Wars from war to peace would enable many more of the U.S.A.'s new immigrant population to participate in the economy. Star Wars is fine for places like North American Rockwell and Livermore Labs, but the rest of the population is turned into fast food clerks peddling burgers to the aerospace workers. An autopoetic economy of direct venture capital for the citizen would cost less than Star Wars, stimulate the economy more effectively, and would be popular enough that citizens would not need to be scared to death to vote for the subsidies it would require.

If one takes all of the five FUTURE TRANSFORMS together, it forms a pattern, a slight akido move that is necessary to transform the militarism of Reagan into a new populist liberalism for a transformed Democratic

Party in the nineties. If the Democratic Party remains the true conservative party of the industrial past, of labor unions and ethnic blocs, it will, like Mondale, become a fossil. And if the Democratic Party tries to become identical to Reagan's party and woo the same constituency, it will only prove itself to be shallow, throughtless, opportunistic, and completly lacking in credibility as well as power. If, on the other hand, a new American ecological party were to try to make it on its own, such a movement on the Left would generate its mirror-opposite on the Far Right, and Lyndon Larouche's thermonuclear fusionists would probably match the Greens vote for vote, with each party taking about 15% of the electorate. It would be far better for the Democratic Party to take the best of the ecological party *and* the best of American Big Science, to move the new ethnic majority in defeat of the white suburban affluent constituency that supports Reagan and Bush. Paradoxically, it is this new Latin and Asian America that is more truly expressive of the Pacific Rim-California culture that first set Reagan into power.

I doubt if the Democratic Party will adopt Gaian politics in 1988; most likely it will try to xerox the Republicans with someone like Iacocca, and our politics will be the typical American cultural situation of Avis and Hertz, Pepsi and Coca-Cola, MacDonalds and Burger King. But history is full of surprises like Chernobyl, and so I would imagine that by 1992, this awful generation of the fifties, these hideous reruns of the anti-intellectual McCarthy era, will have spent themselves out, and just as the sixties introduced a quantum leap in consciousness for the whole human race, so will the nineties take us up one more step. It won't take a national charismatic leader to effect such a cultural shift, for by the nineties the generation of the sixties will be spread throughout the Establishment: as corporation presidents, as politicians, as popular musicians and video artists, as university leaders. As they look around and see themselves in position, they will remember, and those camp followers who now celebrate their neo-conservative orthodoxy will change spots once again to drag out their old sixties' credentials and begin to boast about how many demonstrations, love-ins, and rock festivals they took part in. Once again, it will be fashionable to be idealistic, and patriotic, not simply for Springsteen's U.S.A., but the entire planet. Such is the fantasy of one who came of age in the sixties, and such is my fantasy of a new Gaian form of politics for the nineties.

III. Eight Theses for a Gaia Politique

1. Every intellectual searches for a new ideology, hoping to become another Marx for a better Lenin; but ideology is to the mind what excrement is to the body: the exhausted remains of once living ideas.

2. The Truth cannot be expressed in an ideology, for Truth is the shared life that overlights the conflict of opposed ideologies, much in the same way that the Gaian atmosphere overlights the "conflict" of ocean and continent; therefore, the Truth cannot be "known" by the process of intellectual analysis, critique, or communicative rationality; nor can it be socially administered by a philosophical or religious elite of the best and brightest, be they followers of Mohammed, Marx, Habermas, or E. O. Wilson. Since "knowing" is a form of "false consciousness," elites are institutional reifications of this false consciousness that break up the compassionate feeling of our common life in the world.

3. A World is not an ideology nor a scientific institution, nor is it even a system of ideologies; rather, it is a structure of unconscious relations and symbiotic processes. In these living modes of communication in an ecology, even such irrational aspects as noise, pollution, crime, warfare and evil can serve as constituent elements of integration in which negation is a form of emphasis

and hatred is a form of attraction through which we become what we hate. The Second World War in Europe and the Pacific expressed chaos and destruction *through* maximum social organization; indeed, this extraordinary transnational organization expressed the cultural transition from a civilization organized around literate rationality to a planetary noetic ecosystem in which stress, terrorism, and catastrophes were unconsciously sustained to maintain the historically novel levels of world integration. Through national, thermonuclear terrorism, and, as well, through subnational expressions of terrorism electronically amplified, these levels of stress and catastrophic integration are still at work today. A World should not be seen, therefore, as an organization structured through communicative rationality,[11] but as the cohabitation of incompatible systems by which and through which the forces of mutual rejection serve to integrate the apparently autonomous unities in a meta-domain that is invisible to them but still constituted by their reactive energies. Therefore, ideologies do not map the complete living processes of a World, and unconscious polities emerge independent of "conscious purpose." Shadow economies (such as the drug traffic between Latin America and the United States), and shadow exports (such as the acid rain from the United States to Canada), and shadow integrations (such as the war between the United States and Japan in the forties) all serve to energize the emergence of a biome that is not governed by conscious purpose.[12]

4. Human beings, therefore, never "know" what they are "doing." Since Being, by definition, is greater than knowing, human beings embody a domain structured by opposites by thinking one thing, but doing another; thus negation becomes a form of emphasis in which cops stimulate robbers, celibates stimulate sexuality, and science stimulates irrational superstition and chaos. In the domain of cops and robbers, an interdiction serves to structure a black market and a shadow economy. In the

domain of religious celibacy, an interdiction serves to mythologize repression and energize lust. In the domain of science, the hatred of ambiguity, wildness, and unmanageability creates a superstitious belief in technology as an idol of control and power; thus irrational experiments like nuclear energy and genetic engineering become forms of seemingly managed activity that generate chaos and disease.

5. "Nature" is neither a place nor a state of being; it is a human abstraction that we set up through cultural activities. We then use this abstraction to justify these very cultural activities as "natural." This process of abstraction is an empty tautology. "Nature," in Buddhist terms is groundless; therefore, we cannot appeal to "Nature" to condemn activities as unnatural. As Nature changes with Culture, both are individually empty and linked together in "codependent origination," or *pratityasamutpadha*. Genetic engineering, artificial intelligence, or nuclear power cannot be condemned on the grounds that they are "unnatural"; they can only be rejected on cultural grounds that they are not spiritually wise or aesthetically desirable.

6. The conscious purpose of science is control of Nature; its unconscious effect is disruption and chaos. The emergence of a scientific culture stimulates the destruction of nature, of the biosphere of relationships among plants, animals, and humans that we have called "Nature." The creation of a scientific *culture* requires the creation of a scientific *nature,* but since much of science's activities are unconscious, unrecognizedly irrational, and superstitious, the nature that science summons into being is one of abstract system and concrete chaos, e.g. the world of nuclear power and weapons. The more chaos there is, the more science holds on to abstract systems of control, and the more chaos is engendered. There is no way out of this closed loop through simple rationality, or through the governing systems that derive from this rationalization of society.

7. The transition from one World to another is a catastrophe, in the sense of the catastrophe theory of René Thom. Indeed, a catastrophe is the making conscious of an Unconscious Polity; it is the feeling in Being of a domain that is unknown to thinking. Catastrophes are often stimulated by the failure to feel the emergence of a domain, and so what cannot be felt in the imagination is experienced as embodied sensation in the catastrophe. When rational knowing and political governance no longer serve to feel the actual life of a World, then consciousness becomes embodied in experience outside the world-picture but still within the invisible meta-domain. The conscious process is reflected in the Imagination; the unconscious process is expressed as *karma,* the generation of actions divorced from thinking and alienated from feeling. Catastrophes are discontinuous transitions in Culture-Nature through which knowing has an opening to Being. This moment of passing-together through a catastrophe, this occasion of com-passionate participation, presents an opportunity for a shift from karmic activity to Enlightenment. Thus the transition from one World-Structure to another is characterized by catastrophes in which the Unconscious Polities become visible. At such times there can be a rapid flip-over or reversal in which the unthinkable becomes possible.

8. No governing elite will allow us to think this transition from one World-Structure to another, but imagination and compassion will allow us to feel what we cannot understand. As "Nature" comes to its end in our scientific culture, the relationship between unconscious and conscious will change and the awareness of immanent Mind in bacteria[13] and of autopoesis in devices of Artificial Intelligence[14] will give us a new appreciation of the animism of ancient world-pictures. The "Man" of the historical set of Culture-Nature will come to his end in a new irrational world of angels and devils, elementals and cyborgs. In this science fiction landscape, this

invisible meta-domain in which we already live, the end of Nature as unconscious *karma* makes of Enlightenment and Compassion a new political possibility.

Notes

1. Gregory Bateson, "The Effect of Conscious Purpose on Human Adaptation" in *Steps to an Ecology of Mind* (New York, Ballantine, 1972), pp. 440–448.
2. See Lewis Thomas, "At the Mercy of Our Defenses" in *Earth's Answer: Explorations of Planetary Culture at the Lindisfarne Conferences* (New York, Harper & Row/Lindisfarne, 1977), pp. 156–169.
3. See *The Autobiography of William Carleton* (London, MacGibbon & Kee, 1968), p. 117.
4. I have explained this point in greater detail in my previous book, *Pacific Shift* (San Francisco, Sierra Club Books, 1986), pp. 125–144.
5. Immanuel Wallerstein, *The Modern World-System II: Mercantilism and the Consolidation of the European World Economy, 1600–1750* (New York, Academic Press, 1980), p. 159.
6. Lynn Margulis, *Symbiosis and Cell Evolution* (San Francisco, Freeman, 1981).
7. See Humberto Maturana and Francisco Varela, *Autopoesis and Cognition: The Realization of the Living* (Boston University Studies in the Philosophy of Science, Boston, D. Reidel, 1980).
8. Peter Schwartz, Shell Oil, London, "Address to the E.E.C. Officers for Research and Development," Geneva, June 12, 1986.
9. Francisco Varela and Evan Thompson, *Worlds without Ground: Cognitive Science and Human Experience*, work in progress, probably to appear in 1988.
10. Jacques Attali, *Noise: The Political Economy of Music* (Minneapolis, University of Minnesota Press, 1985), pp. 31, 32.
11. See Jurgen Habermas, *The Theory of Communicative Action*, Vol. One, *Reason and the Rationalization of Society* (Boston, Beacon Press, 1981), p. 397. "If we assume that the

human species maintains itself through the socially coordinated activities of its members and that this coordination has to be established through communication—and in certain central spheres through communication aimed at reaching agreement—then the reproduction of the species also requires satisfying the conditions of a rationality that is inherent in communicative action."

12. See Bateson, "The Effect of Consciousness Purpose on Human Adaptation."
13. See Maurice Panisset and Sorin Sonea, *A New Bacteriology* (Boston, Jones and Bartlett, 1983), p. 8.
14. See Varela and E. Thompson, *Worlds without Ground.*

NOTES ON CONTRIBUTORS

HENRI ATLAN is a professor of biophysics at the University of Paris and at the Hebrew University of Jerusalem, specializing in cellular biology and the theory of self-organization. His books include *L'organization de biologique et la théorie d'information, Entre le cristal et la fumée,* and *A tort et à raison.* He is also a student of Jewish mysticism.

GREGORY BATESON (1904–1980) moved in and out of various disciplines—biology, anthropology, epistemology, linguistics, psychotherapy—and marked each of them with his passage. He did pioneering anthropological studies in New Guinea and Bali, participated in the Macy Foundation meetings that founded the science of cybernetics, developed the double-bind theory of schizophrenia, studied dolphins, and was a Regent of the University of California and a Fellow of the Lindisfarne Association. His books include *Naven, Steps to an Ecology of Mind,* and *Mind and Nature.*

HAZEL HENDERSON is an independent futurist, economic analyst, political activist, and author. Her books include *Creating Alternative Futures* and *The Politics of the Solar Age: Alternatives to Economics.* She is a Fellow of the Lindisfarne Association.

JAMES LOVELOCK is an independent scientist and originator, with Lynn Margulis, of the Gaia Hypothesis. He is the author of *Gaia: A New Look at Life on Earth* and a Fellow of the Lindisfarne Association.

LYNN MARGULIS is a professor of Biology at Boston University and the originator, with Lovelock, of the Gaia Hypothesis. Her books include *The Origin of Eukaryotic Cells, Symbiosis and Cell Evolution, Early Life* and (with Dorian Sagan) *Micro-Cosmos.* She is a Fellow of the Lindisfarne Association.

HUMBERTO MATURANA is a neurobiologist at the University of Chile in Santiago and co-author (with Lettvin and McCulloch) of the seminal paper on the neurobiology of cognition "What the Frog's Eye Tells the Frog's Brain." He is also the co-author, with Varela, of *Autopoesis and Cognition* and *The Tree of Knowledge: A New Look at the Biological Roots of Human Understanding.*

WILLIAM IRWIN THOMPSON is a cultural historian and founder and President of the Lindisfarne Association. He has taught at MIT, York University, and the University of Toronto and has written several books, most recently *Islands Out of Time* and *Pacific Shift.*

JOHN TODD is a co-founder of New Alchemy Institute and is currently President of Ocean Arks International, a not-for-profit research and communication organization. O.A.I. publishes *Annals of Earth* (edited by Nancy Jack Todd), which is available for a minimal contribution of $10 from O.A.I., 10 Shanks Pond Road, Falmouth, MA 02540. Dr. Todd is a Fellow of the Lindisfarne Association.

FRANCISCO VARELA has concentrated on neurobiological and cybernetic mechanisms of cognitive phenomena, especially perception, and the related epistemological issues. His books include *Principles of Biological Autonomy, Autopoesis and Cognition,* and *The Tree of Knowledge* (the latter two written with

Maturana). He currently holds a chair of Cognitive Science and Epistemology at the Ecole Polytechnique in Paris and directs a research laboratory at the Institute of Neuroscience. He is also interested in Buddhist meditation practice and thought, and is a Fellow of the Lindisfarne Association.

1932